THE CHALLENGE OF NUCLEAR-ARMED REGIONAL ADVERSARIES

DAVID OCHMANEK
LOWELL H. SCHWARTZ

D1468722

Prepared for the United States Air Force

PROJECT AIR FORCE

The research described in this report was sponsored by the United States Air Force under Contract FA7014-06-C-0001. Further information may be obtained from the Strategic Planning Division, Directorate of Plans, Hq USAF.

Library of Congress Cataloging-in-Publication Data

Ochmanek, David A.
    The challenge of nuclear-armed regional adversaries / David Ochmanek, Lowell H. Schwartz.
      p. cm.
    Includes bibliographical references.
    ISBN 978-0-8330-4232-3 (pbk. : alk. paper)
      1. Nuclear warfare. 2. United States—Military policy. 3. National security—United States. 4. Strategic forces. 5. International security. 6. Deterrence (Strategy) I. Schwartz, Lowell. II. Title.

    U263.O34 2008
    355.02'17—dc22

                                       2008007737

The RAND Corporation is a nonprofit research organization providing objective analysis and effective solutions that address the challenges facing the public and private sectors around the world. RAND's publications do not necessarily reflect the opinions of its research clients and sponsors.

RAND® is a registered trademark.

Published 2008 by the RAND Corporation
1776 Main Street, P.O. Box 2138, Santa Monica, CA 90407-2138
1200 South Hayes Street, Arlington, VA 22202-5050
4570 Fifth Avenue, Suite 600, Pittsburgh, PA 15213-2665
RAND URL: http://www.rand.org/
To order RAND documents or to obtain additional information, contact
Distribution Services: Telephone: (310) 451-7002;
Fax: (310) 451-6915; Email: order@rand.org

Preface

On October 9, 2006, North Korea tested its first nuclear device. Granted, the explosive yield of the device (estimated at half a kiloton [kt], or the equivalent of 500 tons of TNT) was not, by the standards of most nuclear weapons, impressive. Nevertheless, the fact that an impoverished nation-state, such as North Korea, could develop and test a nuclear device in the face of opposition from the United States and all the other states in northeast Asia is a signal event in international relations. It suggests that the United States and like-minded countries have difficulty denying nuclear weapons to regional powers that seek them. In light of this, prudence dictates that the United States and its allies prepare for the possibility that they might, in the not-too-distant future, confront regional adversaries with deliverable nuclear arsenals.

In anticipation of this future, analysts at RAND have, for some time, been conducting research on the problems that nuclear-armed regional adversaries pose. The insights they have developed raise important questions about U.S. strategy for power-projection operations and about the adequacy of the capabilities that may be available to future U.S. forces. The work documented here, part of a fiscal year 2006 study, "The Future of Deterrence in a Proliferated World," suggests strongly that it would be a mistake to regard nuclear-armed regional adversaries simply as lesser included cases of more powerful adversaries, such as the Soviet Union of the Cold War.

The research documented here was sponsored by the Deputy Chief of Staff, Plans and Programs, Headquarters United States Air

Force and was conducted within the Strategy and Doctrine Program of RAND Project AIR FORCE.

Other RAND Project AIR FORCE documents that address nuclear armament and regional adversaries include the following:

- *Future Roles of U.S. Nuclear Forces: Implications for U.S. Strategy* (MR-1231-AF), by Glenn C. Buchan, David Matonick, Calvin Shipbaugh, Richard Mesic
- *U.S. Regional Deterrence Strategy* (MR-490-A/AF), by Ken Watman, Dean A. Wilkening, Brian Nichiporuk, and John Arquilla
- *Nuclear Deterrence in a Regional Context* (MR-500-A/AF), by Dean A. Wilkening and Ken Watman

RAND Project AIR FORCE

RAND Project AIR FORCE (PAF), a division of the RAND Corporation, is the U.S. Air Force's federally funded research and development center for studies and analyses. PAF provides the Air Force with independent analyses of policy alternatives affecting the development, employment, combat readiness, and support of current and future aerospace forces. Research is conducted in four programs: Aerospace Force Development; Manpower, Personnel, and Training; Resource Management; and Strategy and Doctrine.

Additional information about PAF is available on our Web site: http://www.rand.org/paf/

Contents

Figures

Tables

Summary

The United States, along with other members of the international community, is striving to convince North Korea, Iran, and other states to forgo the development of nuclear weapons. If these efforts do not succeed, the consequences for U.S. and allied security could be profound.

U.S. conventional and nuclear forces will continue to have deterrent effects on the leaders of regional adversary states, such as North Korea and Iran, even if these states field substantial numbers of nuclear weapons. However, defense planners in the United States and elsewhere must begin now to confront the possibility that, in the face of superior U.S. conventional forces, adversaries of this class could see using nuclear weapons to be in their interest under a variety of circumstances during a conflict involving the United States. Several reasons exist for this:

- Regional adversary nations spend only a small fraction of what the United States does on military forces (less than 5 percent in the cases of Iran and North Korea). This virtually guarantees that any serious conflict involving the United States will end in such opponents' defeat if the conflict stays at the conventional level. (See pp. 15–17.)
- Military defeat can have disastrous consequences for authoritarian rulers, who may therefore be prepared to run high risks to stave it off. Facing the prospect of defeat, enemy leaders may perceive that using one or more nuclear weapons may be the most attrac-

tive option open to them if it might deter the United States and its allies from continuing their military operations. (See pp. 36–37.)

- In several conflicts, U.S. forces have demonstrated the capability and will to attack enemy leaders, command-and-control assets, weapons of mass destruction, and delivery means from the outset. Fears of decapitation strikes or disarming counterforce attacks could lead enemy leaders to perceive that they are in a use-or-lose situation, thus heightening the pressure to resort to nuclear use early in a conflict. (See p. 37.)

In short, deterring the use of nuclear weapons by threatening retaliation, which was a mainstay of Cold War military strategy, could be highly problematic in many plausible conflict situations involving nuclear-armed regional adversaries for the simple reason that adversary leaders may not believe that they will personally be any worse off for having used nuclear weapons than if they were to forgo their use. This being the case, U.S. and allied leaders confronting nuclear-armed adversaries will want military capabilities that offer far greater assurance than do today's that adversaries can be *prevented* (as opposed to deterred) from using nuclear weapons. This points to demands for forces that can locate, track, and destroy nuclear weapons and their delivery means before they are launched and, above all, active defenses that can destroy delivery vehicles after they have been launched. Today and for some time to come, the emphasis should be on fielding effective defenses against theater-range missiles, not ICBMs. (See pp. 39–42, 51–52.)

Unless and until highly reliable means of attack prevention become available, U.S. leaders will be compelled to temper their objectives vis-à-vis nuclear-armed regional adversaries, avoiding conflict with them or using military force in limited ways that minimize the adversary's incentives to escalate to nuclear use. (See p. 53.)

Acknowledgments

The work whose results are reported here was one component of a larger project devoted to understanding and preparing for the challenges of nuclear weapons in what many of us have taken to calling the post–post–Cold War world. Our colleagues in the larger enterprise examined the tenets of deterrence theory as they apply to current and future conditions, the changing nature of the "nuclear balance" between the United States and major potential adversaries (Russia and China), and the challenges that might be posed by terrorist groups with one or more nuclear weapons. As we developed and refined our analysis of the challenges posed by nuclear-armed regional state adversaries, our colleagues were generous with their time and expertise, offering suggestions for tightening our logic and clarifying presentation of our ideas. Accordingly, we are indebted to David Shlapak, Jasen Castillo, Forrest Morgan, David Mosher, and Jed Peters. Our colleague Thomas McNaugher also provided helpful comments on the manuscript.

The study of conflicts and crises involving nuclear weapons is unavoidably hamstrung by a paucity of historical data: We have (fortunately) very few historical instances of conflict involving two nuclear-armed adversaries. This reality makes political-military gaming an especially valuable research tool for generating insights about future interactions between nuclear-armed states. Since 1990, our colleagues Roger Molander and Peter Wilson have developed and conducted a series of games called "The Day After . . . ," each of which features nuclear use. We drew heavily on a recent Day After game developed by Peter Wilson using a scenario involving North Korea. We also used the Day After approach to construct our own game featuring a nuclear-armed Iran. Both games have been played many times by a variety of participants, including military officers, civilian officials of the U.S. Department of Defense and other U.S. agencies, as well as analysts and academics. We are grateful to all of these participants for their involvement in the research and the insights they shared with us during and after each iteration of the game.

The authors also wish to express their appreciation to Lawrence Freedman of King's College, London, and to our RAND colleague Robert Levine for their careful and incisive reviews of an earlier draft of this book. Both of these scholars freely lent their expertise on matters relating to nuclear strategy, and the work benefited significantly from their efforts. We are also grateful to an outside reader, Carol Levine, who directed our attention to the potential importance of Pakistan as a nuclear-armed regional adversary, should that nation's internal politics evolve in unfavorable ways. Finally, we are indebted to Cynthia Cook, associate director of PAF, who subjected the draft of this monograph to special reviews to ensure clarity of presentation, and to Lisa Bernard, who expertly edited the manuscript.

Abbreviations

AMRAAM	Advanced Medium-Range Air-to-Air Missile
EMP	electromagnetic pulse
GCC	Cooperation Council for the Arab States of the Gulf, also known as the Gulf Cooperation Council
HEMP	high-altitude electromagnetic pulse
ISI	Inter-Services Intelligence
kt	kiloton
LeT	Lashkar-e-Taiba
ODS	Operation Desert Storm
PSI	pounds per square inch
THAAD	Theater High-Altitude Area Defense, formerly known as Terminal High-Altitude Area Defense
TMD	theater missile defense

CHAPTER ONE

Introduction

A defining feature of the post–Cold War international security environment has been that the United States, acting either alone or with allies and coalition partners, possessed the capability to impose its will on states, such as Serbia and Iraq under Saddam Hussein, that could be termed *regional adversaries*. We define this term to mean countries (1) that pursue policies that are at odds with the interests of the United States and its security partners and that run counter to broadly accepted norms of state behavior and (2) whose size and military forces are not of the first magnitude.[1] The category is useful as a means of distinguishing this group of states from larger, more powerful states, such as Russia, China, and India, which do not share their vulnerabilities to forcible intervention and which, for the present, at least, are pursuing policies vis-à-vis the United States and its allies that are generally more cooperative than confrontational.

The collapse of the Soviet Union was largely responsible for the United States' newfound freedom of action in the 1990s. For one thing, the disappearance of any military threat to NATO Europe freed U.S. conventional forces for missions elsewhere. For another, Moscow's preoccupation with domestic problems and its abandonment of an expansionist ideology meant that regional adversaries of the United States could no longer expect to receive large-scale material assistance or military support from abroad—support that had included, in some cases, Moscow's implicit or explicit nuclear guarantees against U.S. attack.

---

[1] Such countries were once commonly referred to as *rogue states*, but this term, with its inherently value-laden connotations, has fallen out of favor.

Therefore, regional adversaries found themselves isolated, and fears of escalation no longer overshadowed U.S. military engagements in Eurasia.

This state of affairs persists in some cases. The problem is that the regional adversaries likeliest to come into serious conflict with the United States and its regional allies or partners—North Korea and Iran—either have nuclear weapons or have the potential to acquire them.[2] In the near term, these are the two regional-adversary states likeliest to field nuclear weapons. Over the longer term, other plausible adversaries, such as Syria, might join this group, or Pakistan, which already has nuclear weapons, might adopt an adversarial stance if elements hostile to the United States were to take over Pakistan's government.

Nuclear-armed regional powers would present the United States with security challenges that are quite different from those that it faced during the Cold War and in the post–Cold War era. Accordingly, Western strategists will want to understand the ways in which nuclear weapons might affect the behavior of regional adversaries in peacetime, crisis, and conflict, and assess the likely ramifications of this development for U.S. security and defense planning. This book is intended to shed light on these questions.

Chapter Two provides a short primer on the effects of nuclear weapons of the type likeliest to be in the hands of regional

---

[2]   Our research was not predicated on the assumption that either North Korea or Iran was certain to acquire nuclear weapons. Indeed, recent progress and the six-party talks have opened the way to the disabling of at least part of North Korea's nuclear infrastructure. In addition, the U.S. intelligence community judged, in late 2007, that Iran had "halted" the military component of its nuclear-weapon program four years earlier. However, these developments notwithstanding, both countries clearly possess the scientific and engineering capabilities, the economic resources, and at least large parts of the physical infrastructure required for producing nuclear weapons and the means to deliver them. This being the case, prudence demands that military planners consider carefully the potential consequences of nuclear weapons in the hands of these states. On the six-party talks, see Christopher R. Hill, assistant secretary for East Asian and Pacific affairs, "North Korea and the Current Status of Six-Party Agreement," statement before the U.S. House of Representatives Committee on Foreign Affairs, February 28, 2007. On the intelligence community's most recent estimate of Iran's nuclear capabilities, see Office of the Director of National Intelligence and National Intelligence Council, *Iran: Nuclear Intentions and Capabilities*, Washington, D.C., 2007.

adversaries—that is, fission weapons with a yield of between 10 and 20 kilotons (kt). Chapter Three explores the nature of regional adversaries: why they might pursue nuclear weapons and the characteristics of these states that might shape their behavior vis-à-vis the United States and their neighbors. This discussion serves as the foundation for Chapter Four, which examines the strategies and actions that nuclear-armed regional adversaries might undertake, particularly in crisis or conflict situations. Chapter Five concludes the book with a consideration of how the potential for conflict with nuclear-armed regional adversaries might affect U.S. military strategy, operations, and force planning.

The Uniquely Destructive Capabilities of Nuclear Weapons

The fact that nuclear weapons are highly destructive will not come as news to any reader of this book. Therefore, it may seem unnecessary to begin our assessment with a review of the physical effects of nuclear weapons. But any serious consideration of the strategic and operational implications of these weapons should begin with an appreciation for their physical effects. For decisionmakers contemplating the merits and risks of prosecuting attacks on a regional adversary, it will matter whether that adversary's threatened retaliation might kill 1,000, 10,000, or 100,000 people.

This chapter, then, briefly describes what happens when a nuclear weapon with a yield of between 10 and 20 kt—about the size of the bombs dropped on Hiroshima and Nagasaki—is detonated and what effects such blasts might have on military operations and civilian populations. Weapons of this size should be well within the technical capabilities of a regional nuclear power such as North Korea or Iran.

The Effects of Nuclear Weapons

Nuclear weapons have five primary effects:[1] blast, immediate or prompt radiation effects, thermal radiation effects, long-term radioactive fallout, and electromagnetic pulse (EMP). We consider each in turn.

---

[1] This discussion is based on a number of sources, including U.S. Congress, Office of Technology Assessment, *The Effects of Nuclear War*, Washington, D.C., 1979, p. 19; Michael Riordan, ed., *The Day After Midnight: The Effects of Nuclear War*, Palo Alto, Calif.: Cheshire

Blast Effects

Blast effects are caused by the rapid expansion of the fireball produced by a nuclear explosion and the blast wave that results from this expansion. The fireball is caused by the great heat of the explosion, which vaporizes material with which it comes into contact. Essentially everything within the radius of this fireball will be completely destroyed. The heat from the fireball also causes a high-pressure wave to develop and move outward, producing the blast effect. This blast wave is usually measured by the amount of overpressure it produces—that is, the pressure in excess of the normal atmospheric value. Table 2.1 shows the overpressure loads (measured in pounds per square inch, or PSI) produced by nuclear bursts at four different yields and at different ranges from the detonation.[2] It also shows the dynamic pressure, or very high-speed wind, that rushes out from the explosion.

How do these numbers translate into damage? The following general rules can be applied:

- At 20 PSI of static overpressure, even reinforced-concrete buildings are destroyed.
- Ten PSI will collapse most factories and commercial buildings, as well as wood-frame and brick houses.
- Five PSI flattens most houses and lightly constructed commercial and industrial structures.
- Three PSI suffices to blow away the walls of steel-frame buildings.
- Even 1 PSI will produce flying glass and debris sufficient to injure large numbers of people.

Books, 1982; Samuel Glasstone and Philip J. Dolan, *The Effects of Nuclear Weapons*, 3rd ed., Washington, D.C.: U.S. Department of Defense, 1977; D. C. Kephart, *Damage Probability Computer for Point Targets with P and Q Vulnerability Numbers*, Santa Monica, Calif.: RAND Corporation, R-1380-1-PR, 1977; and Chuck Hansen, *US Nuclear Weapons: The Secret History*, Arlington, Tex.: Aerofax, 1988.

[2] The two smallest yields selected for the table represent a reasonable range for purely fission ("atomic") weapons of the kinds likeliest to be available to regional adversaries or a terrorist group. The 1-megaton yield is representative of a fairly large thermonuclear ("hydrogen") warhead.

Table 2.1
Overpressure and Dynamic Pressure as a Function of Yield and Distance for Airbursts

Yield (kt)	Distance (miles)	Overpressure (PSI)	Wind (MPH)
5	0.2	17	400
	0.5	9	250
	1.0	3	100
	2.0	1	40
	5.0	—	—
20	0.2	80	1,500
	0.5	19	425
	1.0	6	200
	2.0	2	80
	5.0	—	—
100	0.2	>200	>2,000
	0.5	38	1,500
	1.0	16	350
	2.0	5	160
	5.0	1	45
1,000	0.2	>200	>2,000
	0.5	200	2,000
	1.0	42	800
	2.0	17	380
	5.0	4	130

SOURCE: Glasstone and Dolan, 1977, pp. 96–102.

NOTE: The data shown here apply to airbursts at optimum height above ground level. See Glasstone and Dolan, 1977, pp. 96–102.

Prompt Radiation Effects

Prompt radiation refers to radiation emitted within one minute after the explosion. It consists primarily of high-energy gamma rays and neutrons. People close to ground zero may receive lethal doses of radiation; however, they will most often be killed by the blast wave and thermal pulse. In typical nuclear attack, only a relatively small proportion of deaths and injuries will result from initial radiation.

Thermal Radiation Effects

Thermal radiation causes flash burns to the skin of exposed individuals. The clothes of exposed individuals can catch on fire, and many will receive severe second-degree burns. The thermal effect could also ignite flammable materials at substantial distances. Damaged buildings around the 5-PSI ring have the potential to ignite, causing widespread damage if the fires are not contained. In urban environments, it is also likely that fires will occur from damage to systems for distributing gas and electricity.

Long-Term Radioactive Fallout

Fallout particles are produced through the interaction of radioactive elements of the weapon with soil, water, and other materials in the vicinity of the explosion. These particles may be dispersed over large areas downwind, and their effects can be felt at distances well beyond the range of other effects of the nuclear explosion. This fallout starts to deposit within 10 to 15 minutes after the detonation in the area and can continue to spread for the next 24 to 48 hours. Areas contaminated by fallout can remain uninhabitable for many years following a nuclear detonation. The direction, intensity, and dispersal of fallout are highly dependent on local conditions and cannot be easily predicted.

Electromagnetic Pulse

This results from secondary reactions occurring when gamma radiation from a nuclear detonation is absorbed into the air or ground. It causes powerful surges of electrical and magnetic energy that can damage or disable exposed electronic equipment.

The relative importance of each of these mechanisms varies depending on the size of the weapon, its mode of detonation, prevailing weather conditions, and other factors. Importantly, the intensity of these effects at various distances from the blast varies according to whether the weapon is exploded in the air (a so-called "airburst") or on the surface (a "ground burst"). Generally speaking, the blast and other immediate effects of ground-burst weapons will be especially intense in the immediate vicinity of the detonation but will fall off comparatively rapidly as the distance from ground zero increases. A ground-burst weapon will also create a crater, the debris from which is energized and made radioactive in the mushroom cloud and becomes fallout. An airburst spreads the immediate effects—blast, radiation, heat, and EMP—out more evenly but, because its fireball typically does not touch the ground, produces little fallout.

Effects of Three Types of Nuclear Attacks

Table 2.1 shows that, even for the smallest weapon, individuals in the open a mile from the blast will be in danger of death or severe injury. A 20-kt weapon—a size we would expect to make up the arsenal of a first-generation nuclear-armed adversary such as North Korea or Iran—will knock down buildings a mile from the blast and seriously injure people 2 miles away.

As a point of reference, the National Mall in Washington, D.C., is roughly 2 miles long, stretching from the U.S. Capitol on its eastern end to the Lincoln Memorial overlooking the Potomac River. Figure 2.1 shows this area, overlaid with the approximately 5-PSI ring—the area that would be subjected to at least 5 PSI of overpressure and winds greater than 200 miles per hour—from a 20-kt airburst. It shows that much of official Washington would be obliterated by a single such weapon. The Capitol, White House, and many cabinet department buildings would be destroyed or heavily damaged. Tens, perhaps hundreds of thousands of people, would be killed and injured.[3]

The urban landscape can affect the propagation of effects from very low-altitude detonations or ground bursts. Blast and radiation will propagate freely up broad avenues or across open spaces but can be attenuated by blocks of substantial buildings, such as those in the "skyscraper canyons" of New York City. Because of these uncertainties, precisely estimating the numbers of casualties or the extent of damage that might result from a given detonation can be difficult. Prior RAND research assessed the number of deaths and injuries potentially caused by a 10-kt terrorist bomb detonated at street level at the noon hour in midtown Manhattan. Taking into account the effects of the city environment on the spread of the blast wave, about 90,000 people would likely die and another 400,000 would be injured.[4]

Should North Korea or Iran succeed in acquiring a small arsenal of nuclear weapons, either state could threaten the cities and economies of important U.S. allies. Simple calculations suggest that a single 20-kt airburst over central Tokyo could kill upward of 140,000 people. The same attack on Seoul—where population densities are, on average, about 25 percent higher than they are in Tokyo—could result in nearly 185,000 deaths.[5]

---

[3] It is worth noting that even a "small" 5-kt weapon would blanket the circle depicted in Figure 2.1 with upward of 3 PSI and subject it to winds of 100 MPH or greater, causing substantial damage, death, and injury.

[4] Estimated using the approximation that "everyone inside 5 PSI dies; everyone outside lives" and U.S. Department of Transportation estimates of a workday population density in midtown Manhattan of 167,100 people per square mile. See U.S. Congress, 1979, p. 19, and U.S. Department of Transportation, Federal Transit Administration, "Long Island Rail Road Access to Manhattan's East Side (East Side Access), New York, New York," November 1999.

[5] Calculations again performed using the 5-PSI cookie cutter approach. Because of the airburst nature of the attack, the blast is assumed to propagate freely, resulting in higher overpressures over a larger area than in the New York attack described earlier. Central Tokyo—made up of 23 *ku* (wards)—has an average population density of about 35,000 persons per square mile, according to the Tokyo Metropolitan Government ("Tokyo's Geography, History and Population," undated Web page). The average density for Seoul in 1999 was 44,191 persons per square mile, according to Demographia ("Seoul: City Population, Area and Density by Administrative District," undated Web page). While not specified in

Figure 2.1
Area of at Least 5 PSI: 20-kt Airburst over the National Mall

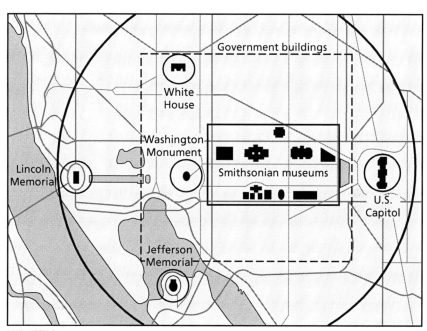

RAND *MG671-2.1*

It is possible a regional nuclear power would attempt a different type of nuclear attack. Instead of attacking a city and causing massive civilian causalities, the adversary might attack a military airbase or an important seaport. Figure 2.2 illustrates the range of serious blast effects from a 20-kt airburst over Osan Air Base in South Korea.

The circled area encompasses the parts of the base that would be subjected to 5 PSI overpressure or more. If accurately delivered, a single fission weapon would cause severe damage to most above-ground structures on the base that were not specially hardened (e.g., aircraft shelters). Any exposed vehicles on the base, including aircraft, would be destroyed. If the blast were a ground burst, it would likely produce

the sources, these appear to be residential densities; the comparable figure for New York is a little less than 26,000. Obviously, a midday attack on a bustling business district—which is the scenario we discuss for Manhattan—could produce several times as many casualties.

Figure 2.2
Area of at Least 5 PSI: 20-kt Airburst over Osan Air Base, South Korea

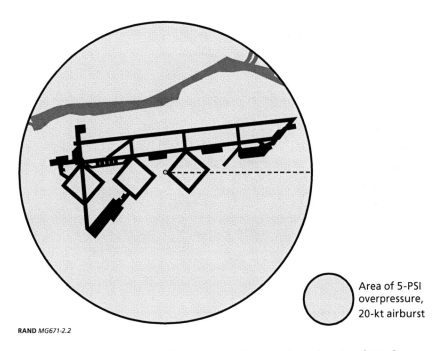

Area of 5-PSI overpressure, 20-kt airburst

a crater approximately 140 feet in radius with a depth of 70 feet.[6] A crater of this size near the center of the main runway could completely shut down air operations for a substantial period. Depending on how much warning time was given prior to the detonation, 20 percent or more of the military personnel on the base could be killed, with most of the remainder being wounded. The local population around the base would also be affected, with the potential for thousands of civilian fatalities and tens of thousands with severe injuries.

A high-altitude EMP (HEMP) attack is also an option. This would require producing a nuclear explosion at least 40 km above the earth's surface and would interfere with electronic equipment by causing physical damage or temporary impairment to electronic com-

---

[6] Glasstone and Dolan, 1977, p. 253.

ponents. However, due to its high altitude, the nuclear device would cause minimal direct damage on the earth's surface.

The precise effects of a HEMP on a particular electronic system are difficult to predict. Only limited experimental data on the effects of a HEMP blast are available because no high-altitude tests have been undertaken since the Limited Test Ban Treaty was signed in 1963.[7] However, analysis suggests that an EMP detonation has the potential to damage all unprotected electronic equipment within line of sight of the explosion. In general, EMP can disable computer networks and critical infrastructure supporting power and communications. In addition, EMP can penetrate the earth several feet to affect underground cables, though damage to these tends to be less severe. Unhardened electronics in aircraft are also susceptible. A HEMP attack might hamper military operations and certainly would have serious effects on key civilian infrastructures, such as power grids and telecommunication networks.[8]

---

[7] Governments of the United States of America, the United Kingdom of Great Britain and Northern Ireland, and the Union of Soviet Socialist Republics, *Treaty Banning Nuclear Weapon Tests in the Atmosphere, in Outer Space and Under Water*, August 5, 1963.

[8] For a detailed examination of EMP, see John S. Foster Jr., Earl Gjelde, William R. Graham, Robert J. Hermann, Henry M. Kluepfel, Richard L. Lawson, Gordon K. Soper, Lowell L. Wood Jr., and Joan B. Woodard, *Report of the Commission to Assess the Threat to the United States from Electromagnetic Pulse (EMP) Attack*, Vol. 1: *Executive Report*, 2004.

CHAPTER THREE

Characteristics of Nuclear-Armed Regional Adversaries

What makes nuclear-armed regional adversaries distinctive from other state adversaries? We begin to address this question by considering the motivations for regional adversaries' pursuit of nuclear weapons. Nuclear weapons may be seen as serving a number of purposes. Iran, for example, is thought to be pursuing them for a combination of reasons:[1]

- to deter military threats or attacks by the United States and, perhaps, others
- to redress its military inferiority vis-à-vis Israel, Pakistan, India, and Russia—neighboring states that have nuclear weapons
- to enhance national prestige and influence
- to shore up domestic political support
- to ensure the survival of the regime in the event of war.

The North Korean regime undoubtedly shares most of these motivations. It might also see its nuclear program as a source of leverage on the United States, Japan, South Korea, and China for extracting economic assistance.

For authoritarian or despotic leaders, such as North Korea's Kim Jong Il, deterring threats to the survival of the regime may be the most compelling motivation for going nuclear. Such leaders historically have

---

[1] See Judith Share Yaphe and Charles D. Lutes, *Reassessing the Implications of a Nuclear-Armed Iran*, Washington, D.C.: Institute for National Security Studies, National Defense University, McNair paper 69, 2005, pp. 3–5.

exhibited a preoccupation with the survival of their regimes. This is due, in part, to the problem of a deficit of legitimacy that typically afflicts such regimes. It is also due to the fact that, in situations in which power is seized and held forcibly, the survival of the regime is often synonymous with the personal survival of those at the top of the regime. Throughout history, many dictators have reached the pinnacle of national power only by ruthlessly eliminating and intimidating rivals. This being the case, a change in leadership in a state such as North Korea is often brought about by the forcible overthrow of the incumbent and can be the occasion for a long-delayed settling of scores.

The situation in Iran is more complicated. Iran has both democratic and authoritarian elements within its governing structure. It has a popularly elected president and a parliament whose membership is shaped by the clerical establishment but also by the electorate. At the same time, the nonelected, religious establishment commands ultimate authority in the country. Deep and enduring philosophical differences divide Iran's governing elites into three camps: hardliners, pragmatists, and reformers. And many elements within Iranian society are known to be dissatisfied with the regime's performance. Yet the regime has shown a great deal of resiliency. Factions within the regime itself and in the society writ large tend to close ranks when confronted with pressure or threats from external sources.[2]

Notwithstanding these differences between North Korea and Iran, regime survival would be a core objective for both nations in any crisis or conflict.[3] To leaders concerned with their ability to maintain a grip on power in the event of war, the value of nuclear weapons is obvious: If an attack by a U.S.-led coalition would pose a significant threat

---

[2]   Ray Takeyh, *Hidden Iran: Paradox and Power in the Islamic Republic*, New York: Henry Holt, 2007, pp. 29–40.

[3]   The behavior of Saddam Hussein and Slobodan Milosevic during their confrontations with the United States and its coalition partners was consistent with this core objective. See Stephen T. Hosmer, *The Conflict Over Kosovo: Why Milosevic Decided to Settle When He Did*, Santa Monica, Calif.: RAND Corporation, MR-1351-AF, 2001; and Stephen T. Hosmer, *Why the Iraqi Resistance to the Coalition Invasion Was So Weak*, Santa Monica, Calif.: RAND Corporation, MG-544-AF, 2007.

to your regime and your nation cannot afford conventional forces capable of deterring or defeating such an attack, you may regard nuclear weapons as the answer.

For many authoritarian leaders, the prospect of the United States forcibly overthrowing them is not an abstract proposition. Both of the U.S. national security strategy documents released by George W. Bush declared that the ultimate goal of the United States is "ending tyranny."[4] And both North Korea and Iran are cited as examples of the types of regimes about which the United States harbors grave concerns.[5] The overthrow of the Taliban in Afghanistan in 2001 and Saddam Hussein in 2003 probably heightened the determination of other regional adversaries to find a means of fending off such attacks.

Key Characteristics of Nuclear-Armed Regional Adversaries

It would be a mistake to think of regional adversaries with nuclear weapons simply as smaller, weaker versions of the nuclear-armed states with which the United States has had a long-standing deterrent relationship—namely, Russia (and before that, the Soviet Union) and China. Regional adversaries have several characteristics, addressed below, that distinguish them from larger and more powerful adversaries. These differences underlie our finding that a strategy and set of supporting capabilities different from those that served in the Cold War will be called for to deal with nuclear-armed regional adversaries.

Inferior Conventional Forces

The leaders of regional adversary states recognize that their military forces are locked into a position of marked inferiority vis-à-vis U.S. conventional forces. Given constraints on their material and human

---

[4]   George W. Bush, *The National Security Strategy of the United States of America*, Washington, D.C.: Executive Office of the President, 2002; George W. Bush, *The National Security Strategy of the United States of America*, Washington, D.C.: White House, 2006.

[5]   See Bush, 2006, p. 1.

resources, such states can, at best, hope to effectively challenge U.S. expeditionary forces in one or two areas of conventional military capability, such as mine warfare or air defense. They may also choose to confront the United States with the prospect of "irregular" challenges, such as guerrilla warfare, insurgency, and terrorist attacks, as part of their response to a potential U.S. attack. What is beyond their means is the ability to counter U.S. theater forces across the board.[6]

This does not mean that an invasion of these states or other types of military action would necessarily be quick or low-cost operations for the United States. In the case of Iran, for instance, the country's sheer size would pose very serious challenges to an invading force that intended to invest the capital and occupy most of its territory. If a substantial portion of the population were mobilized to oppose an occupying force, that force could face difficulties far more daunting than those that coalition forces have faced in Iraq since the fall of Saddam Hussein's regime. And if the United States were to launch a punitive air attack against Iran, the Iranians might be able to counter that with terrorist attacks against U.S. interests in the region or elsewhere.

Nevertheless, the leaders of adversary states understand that, if military operations are confined to the conventional level, they cannot keep large-scale U.S. expeditionary forces from deploying to their regions and from operating at a fairly high tempo. Nor can they prevent U.S. forces from destroying a wide range of high-value political, economic, and military assets. In the absence of costly and sophisticated conventional weapons that could constitute an effective antiaccess capability, it is clear that the adversary regime's fielded forces, as well as its command and control communications and perhaps its ability to control its population, would be badly damaged from the open-

---

[6]   This is quite different from the situation of major powers, such as China. As China continues the rapid modernization of its armed forces, its prospects for confronting the United States with a viable conventional deterrent capability are improving. Recent Chinese investments in a variety of antiaccess capabilities suggest that China is preparing a range of conventional counters to U.S. power-projection capabilities. See Roger Cliff, Mark Burles, Michael S. Chase, Derek Eaton, and Kevin L. Pollpeter, *Entering the Dragon's Lair: Chinese Antiaccess Strategies and Their Implications for the United States*, Santa Monica, Calif.: RAND Corporation, MG-524-AF, 2007.

ing days of a conflict with the United States. Finding an affordable means of deterring or blunting such an attack, then, must be a prime concern of these regimes.

Small but Survivable Nuclear Forces

The wherewithal to develop and build nuclear weapons is getting easier to come by, but it is still expensive. For example, the North Korean nuclear reactor at Yongbyan, which has been the centerpiece of that country's nuclear program, is thought to be capable of turning out approximately 6 kg of weapon-grade plutonium per year. After undergoing reprocessing, this quantity of plutonium is sufficient to make a single fission weapon with a yield between 10 and 15 kt.[7] North Korea is also suspected of harboring a secret program to produce highly enriched uranium using centrifuges. This infrastructure might be sufficient for producing perhaps two or more weapons per year.[8] Iran's nuclear infrastructure, which also incorporates a reactor for producing plutonium and an unknown number of centrifuges, might be of a similar scale.[9] Of course, both countries may have built other facilities that would add to this capacity, but the point is that we should not expect to face regional adversaries armed with hundreds of nuclear weapons or with very powerful fusion weapons, at least for the coming decade or perhaps longer. Rather, should diplomacy fail, over the next 10 years or so, it seems reasonable to assume that these adversaries could field between one dozen and three dozen fission weapons.

Note also that generating fissile material and even testing an explosive nuclear device is not the same thing as having a reliable, deliverable weapon. As adversary states make the transition to nuclear-armed adversaries, there will be a period of ambiguity during which we

---

[7]   Larry A. Niksch, *North Korea's Nuclear Weapons Program*, Washington, D.C.: Congressional Research Service, Library of Congress, IB91141I, August 31, 2005, pp. 11–12.

[8]   Niksch, 2005, pp. 11–12.

[9]   See David Albright and Corey Hinderstein, "Iran: Countdown to Showdown—The International Community Has Given Iran Until November to Come Clean," *Bulletin of Atomic Scientists*, Vol. 60, No. 6, November–December 2004, pp. 67–73.

(and perhaps they) are unclear whether they have weapons that could be used with high confidence in a conflict.

Perhaps more important than the size of the adversary state's nuclear arsenal are its posture and survivability. A small arsenal deployed in a way that was vulnerable to detection and attack would greatly reduce, rather than strengthen, an adversary state's security. Recognizing this, states such as North Korea and Iran have presumably taken pains to disperse and hide key components of their nuclear programs and arsenals. North Korea has long been notorious for its propensity to build facilities deep underground. Given this, the judgment of most experts who have assessed these programs is that even a large-scale air attack could, at best, set its nuclear programs back somewhat, for example, by destroying above-ground facilities such as plutonium-producing reactors and reprocessing facilities.[10] But we cannot assume that U.S. forces have the ability to prevent a regional power from developing nuclear weapons, nor can they destroy or neutralize carefully deployed arsenals short of invading and occupying the enemy's country.[11]

Limited Delivery Options

For the near to midterm, regional adversaries armed with nuclear weapons will not have the capability to deliver those weapons to the United States using "standard" means, i.e., long-range missiles or mili-

---

[10] One expert noted the following:

> U.S. military strikes could probably destroy North Korea's future ability to produce and reprocess plutonium for use in nuclear weapons . . . . These strikes could potentially remove North Korea's ability to produce large quantities of plutonium for the next several years. However an attack is highly unlikely to destroy any existing North Korean nuclear weapons capability. Because the facilities involved in North Korea's uranium enrichment program have not been located (and are likely in hardened or underground sites that are difficult to destroy), military strikes would be unable to prevent North Korea from producing fissile material via uranium enrichment."

See Phillip C. Saunders, "Military Options for Dealing with North Korea's Nuclear Program," Center for Nonproliferation Studies, James Martin Center for Nonproliferation Studies, January 27, 2003.

[11] Of course, even a full-scale invasion might well not serve, since it could provoke the attack that it was intended to prevent.

tary aircraft. Neither Iran nor North Korea possesses intercontinental-range bombers, and the missiles that they have actually deployed to date are assessed to have a maximum range (see Figure 3.1), carrying a plausible nuclear payload, of 1,300 km.[12] North Korea is developing the longer-range Taepo Dong 2 missile which, though highly inaccurate, is thought to be capable of delivering a nuclear warhead to Alaska or Hawaii. North Korea has also been working to extend the range of its missiles in order to be able to reach the west coast of the

Figure 3.1
Maximum Ranges of Operational Iranian and North Korean Ballistic Missiles, c. 2010

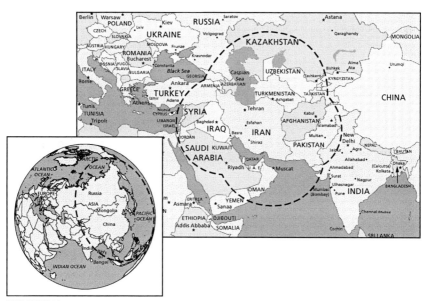

RAND *MG671-3.1*

---

[12] North Korea's No Dong medium-range ballistic missile, which is operational, is believed to have a maximum range of 1,300 km. Iran's Shahab-3 also has an estimated range of 1,300 km. The Iranians are developing the Shahab-3A, which is expected to have a maximum range of up to 1,800 km. William S. Cohen, *Proliferation: Threat and Response*, Washington, D.C.: Office of the Secretary of Defense, 2001, pp. 12, 37. See also MissileThreat, "No-Dong 2," undated Web page (a); and MissileThreat, "Shahab-3," undated Web page (b).

continental United States.[13] The U.S. national missile defense system has been designed to be able to shoot down a modest number of such missiles. So the problem of nuclear weapons in the hands of regional adversaries devolves, for the near term, at least, largely to one of threats to targets in the theater or region of conflict and to the threat that one or more weapons could be delivered to the United States via covert or unconventional means.

For U.S. adversaries seeking to threaten targets within their regions, short- and medium-range ballistic missiles provide the most effective delivery means. When deployed on mobile launchers, these weapons have proven to be highly survivable against air attacks. Moreover, despite some progress in the development of "hit-to-kill" missile defenses, warheads delivered by ballistic missiles have a substantial probability of reaching their targets. Point defense systems, such as Patriot, have small footprints, making it impractical to defend large populated areas with such systems. And any "thin" deployment of missile defenses can be overwhelmed by modest-sized (between 10 and 20 missiles) salvo attacks, which are well within the capabilities of U.S. regional adversaries. Regional adversaries might also seek to employ manned aircraft or cruise missiles as nuclear delivery vehicles, though U.S. forces have shown themselves to be quite adept at defeating attacks by enemy aircraft.

In the absence of a robust intercontinental ballistic missile force, attempting to hold at risk targets within the United States will be tricky. Adversary states might find it necessary to rely on "slow-motion" means of attack, such as a cargo ship, which could conceal one or more nuclear weapons in its hold or, perhaps, launch a short-range missile from a position offshore. Or a civilian airliner might be used as a delivery platform. Both approaches would cause difficulties for U.S. defenders because of the difficulty in sorting out a potential threat from the hundreds or thousands of legitimate vessels and aircraft that approach the United States every day. But both approaches also introduce some complications from the attacker's perspective. Most obviously, they introduce independent actors and other sources of uncertainty into the

---

[13] MissileThreat, undated (c).

sequence of events between command and weapon detonation, all of which reduce to one degree or another the probability of one's orders being carried out effectively.[14]

Targeting Options

Regional adversaries considering potential targets within their regions will have an array of options available to them. The most potent deterrent option available to a regional adversary may be to threaten to attack major cities or vital economic assets with one or more nuclear weapons. As Chapter Two makes clear, a single fission weapon detonated at low altitude over a major city, such as Seoul or Tokyo, could cause well over 100,000 prompt fatalities. A similar attack on the oil export facilities at Dhahran could severely damage the infrastructure over an area of several square miles.

The possibility of incurring damage on this scale would, to say the least, give any decisionmaker pause. But the very destructiveness of such attacks would also make them highly risky for the regional adversary: The adversary would have "killed the hostage" and perhaps held back very little with which to deter a truly devastating retaliation from the United States. Accordingly, we must assume that regional adversaries will also consider using their nuclear weapons to threaten or undertake less consequential attacks. Table 3.1 lists a variety of ways in which a regional nuclear power with a dozen or so deliverable nuclear weapons could use those weapons without resorting to direct nuclear attacks on major cities.

An adversary might, for example, threaten or attack bases used by the air forces of a neighboring state or the United States, perhaps focusing on those far removed from population centers. Or they might attempt to attack concentrations of U.S. or allied ground forces in garrisons or in the field. Alternatively, they might elect to detonate

---

[14] An extreme version of deterrence by the threat of slow-motion or nonstandard means of attack might be called the *dandelion strategy*. A regional leader might seek to deter U.S. military operations against his or her state by threatening to give nuclear weapons to anti-U.S. terrorists, who would then disperse with them to covert locations outside of the adversary country.

Table 3.1
Potential Nuclear Use Options for Regional Adversaries

Objective	Action	Employment Option
Warning	Nuclear demonstration or test	Underground nuclear test Above-ground nuclear test Above-ground nuclear demonstration over adversary's territory (no damage)
Counterforce	Nuclear detonation to disrupt or damage adversary's military forces	EMP blast above air bases EMP blast above naval forces Detonation upwind from air base causing light fallout over base Direct attack on an air base Direct attack on ground forces
Countervalue	Nuclear detonation to damage adversary's civilian infrastructure	Detonation upwind of capital city causing light fallout EMP blast over capital city

a weapon at high altitude so that the EMP from the detonation disabled electronic systems over a wide area but no damage was caused by blast, overpressure, fire, or radiation. Such threats or attacks might be intended to show resolve and thus dissuade opponents from prosecuting military operations against the adversary; failing this, they would, if effective, reduce the capabilities of the forces brought to bear against the adversary.

Clashes of Interests

Another important characteristic of regional adversaries is so obvious that it is easy to overlook. That is that they are, in fact, adversaries: They pursue international objectives that are, in important ways, antithetical to those of the United States. The sources of these conflicting objectives, of course, vary from case to case. Often, they are rooted, to some degree, in the nature of the regime itself and its claim to power, as described below. If a nation's leadership seeks to shore up its legitimacy by defining itself in terms of its opposition to important elements of the international system (as, for example, the Soviets and other communist nations did), it is almost axiomatic that such a nation will pursue policies that most other nations will find objectionable.

For reasons best known to themselves, the leaders of North Korea have consistently espoused as their goal the unification of the peninsula under their rule, by force, if necessary. North Korea is, by its own choice, diplomatically and economically isolated from much of the outside world. Kim Il Sung's economic strategy of self-reliance has been a spectacular failure, with an estimated 2 million to 3 million people—more than 10 percent of the population—starving to death there during the 1990s.[15] And as their nation's economy has fallen further and further behind those of the rest of East Asia, North Korea's leaders have found it expedient to engage in drug trafficking, counterfeiting, and the sale of missiles and nuclear technologies as ways of raising money. Such a regime must have strong reasons to be dissatisfied with and perhaps alarmed by the status quo. As the leadership in Pyongyang compares its situation to that of its counterparts in South Korea, it can hardly fail to see that trends are not, by its lights, favorable.

Iran's place in the international community is quite different from North Korea's. Iran aspires to become the dominant actor in the Persian Gulf region and in the broader Middle East. Its pursuit of this goal is made feasible by its geographic size and demographic weight and by its control over very sizable oil reserves.[16] Judging by their recent statements, some political leaders in Iran today are animated by a degree of revolutionary fervor, which, on several occasions, has been expressed in extreme rhetoric toward Israel and the West.[17] Iran, as much as any

---

[15] Andrew S. Natsios, *The Politics of Famine in North Korea*, Washington, D.C.: U.S. Institute of Peace, 1999.

[16] Iran's population of 68 million makes it roughly 2.5 times larger than either of its closest challengers in the Persian Gulf region—Iraq and Saudi Arabia. Overall, Iranians constitute 52 percent of the population of the entire Persian Gulf region. Iran's proven reserves of 133 billion barrels of crude oil constitute nearly 20 percent of the known reserves in the Persian Gulf region. See U.S. Central Intelligence Agency, "Iran," *The World Factbook*, Washington, D.C., ongoing.

[17] For example, in October 2005, Iran's President Mahmud Ahmadi-Nejad stated, in an address to a conference in Tehran titled "The World Without Zionism," that "Israel should be wiped off the map." In December, Iran's Supreme Leader, Ali Hoseini-Khamenei, said in a meeting with the head of Hamas's politburo that the only way for the Palestinians to liberate their land was through armed struggle. See Al Jazeera, "Ahmadinejad: Wipe Israel Off Map,"

other state, has embraced the political agenda of radical Islam, and it seeks to overturn, or at least substantially alter, the international order in the Middle East, as upheld by the United States. Iran has provided sustained and substantial support to armed Shia factions in Iraq, Lebanon, and Palestine in an effort to extend its influence. Accordingly, it may well be that, absent a change in its strategic orientation, a nuclear-armed Iran would seek to advance its revisionist agenda more aggressively than it has heretofore, perhaps by conducting terrorist operations and other forms of violence below the level of large-scale warfare.[18]

In short, given the past behavior and current policies of both North Korea and Iran, it is not difficult to conceive of ways in which the United States and its security partners could find themselves contemplating military operations against either country. And for different reasons—Kim Jong Il out of a sense that he has little to lose and Iran's leadership out of nationalist ambition fueled by religious-revolutionary zeal—both countries may be willing to accept a great deal of risk once conflict breaks out.

By contrast, neither Russia nor China today seeks to challenge the foundations of the international order as the Soviet Union and Maoist China did. Leaders of both countries seem to have concluded that the current international environment is favorable for their economic development and that neither the United States nor its allies poses an active threat to their political systems. Too, both Russia and China, by virtue of their economic assets and diplomatic clout, have more attractive and less risky means for advancing their national interests than by threatening or unleashing aggression. In light of this, one would expect the leaders of both countries to behave in fairly risk-averse ways: One would not expect them to go about seeking to foment crises and, to the

---

October 28, 2005. See also Pepe Escobar, "But It's So Cold in Alaska," *Asia Times Online*, December 16, 2005.

[18] Iran's sponsorship of the terrorist attack on Khobar Towers in Dhahran in 1996, which killed 19 Americans, has been fairly well established. In July 2001, the U.S. Department of Justice issued an indictment of 14 men on charges of murder and conspiracy for the bombing. The indictment alleged that all 14 were members of the Islamic militant group Hizballah and that this group received support from individuals within the Iranian government. See "Khobar Towers Indictments Returned," *CNN.com*, June 22, 2001.

extent that military power figures in their relations with the United States and its allies, it is very much in the background.[19]

Regimes of Questionable Legitimacy and Stability

Finally, it is possible that the nature of the North Korean and Iranian regimes themselves affects their behavior and propensity to take risks. As is well known, Kim Jong Il, who inherited his position atop the totalitarian North Korean regime from his father, the "Great Leader," rules through some combination of ideological mobilization, personality cult, bribery, and sheer terror. Perhaps the best indicator of the relative importance of these factors is Kim's allocation of resources. Like most dictators, he puts very considerable resources into organs of state security that are devoted to the detection and suppression of internal dissent. While it is perilous to rest too much of one's analysis on speculations about the calculations of Kim Jong Il, he must be aware of the extent to which his regime has failed to provide the vast majority of its people with any tangible reasons to support it other than fear. The existence next door of a vibrant South Korea must be particularly vexatious to him, since the people of North Korea have been fed a steady diet of propaganda since the late 1940s about how they live in a workers' paradise while their brethren to the south are enslaved by a ruthless capitalist elite. Judging from the vehemence with which the regime regularly lashes out at its favorite foreign enemies, it must find that fomenting hostile relations with foreign powers is helpful in keeping the ideological embers alive at home.[20]

---

[19] The worm in the apple here, of course, is Taiwan, which could become a casus belli between China and the United States. If this issue can be managed successfully, however, the prospects are favorable for the international system to adjust peacefully to China's growing power and influence.

[20] The following is one example of a typical pronouncement from Pyongyang, broadcast on September 20, 2005: "Joint military exercises in south Korea . . . should be bitterly denounced by the whole world as they were the ones for aggression staged by the hateful U.S. imperialists and the sycophantic and treacherous forces of south Korea to invade [North Korea]." See Korean Central News Agency of the Democratic People's Republic of Korea, "U.S. Imperialists' Anti-DPRK Moves Denounced," September 20, 2005.

Hard-liners within the government and the clerical establishment in Iran likewise know that they are deeply unpopular with many of their compatriots, particularly urban youth, educated professionals, and large elements of the country's business elite and merchant class. One public-opinion survey conducted in Iran before the 2005 Iranian presidential election there indicated that 66 percent of respondents supported policies to "reform" the government, 23 percent wanted "radical change," and just 11 percent thought that the existing political system and balance of power were acceptable.[21] This has led Iran's conservative clerics to exercise, on a large scale, their prerogative, under the revolutionary constitution, of disqualifying liberal and reform-minded candidates from elections. It has also forced them to rely from time to time on physical means of repression.

In short, paradoxically, the brittleness of dictatorial or authoritarian regimes can cause them to act somewhat recklessly. The weaker the regime perceives itself to be at home, the likelier it may be to take risks abroad. As Watman et al. observed in their assessment of regional deterrence strategies in the mid-1990s,

> [A]lthough the proximate cause of an international crisis involving a [totalitarian or authoritarian] state may be an external event, its deeper causes are often more a function of domestic threats to the weightiest interests of the leadership. The stakes could not be higher for these regimes, and they behave accordingly.[22]

Slobodan Milosevic is a prominent, recent example of a leader who sought conflict and confrontation as a means of maintaining support for his position in power (see Box 3.1).

---

[21] Nazgol Ashouri, "Polling in Iran: Surprising Questions," *PolicyWatch*, Vol. 757, May 14, 2003.

[22] Ken Watman, Dean A. Wilkening, Brian Nichiporuk, and John Arquilla, *U.S. Regional Deterrence Strategy*, Santa Monica, Calif.: RAND Corporation, MR-490-A/AF, 1995, p. 35.

Box 3.1
Slobodan Milosevic: An Instigator and Exploiter of Crises

Slobodan Milosevic, who dominated political life in Yugoslavia through a combination of authoritarianism and demagoguery from 1991 until 2000, provides a clear example of a national leader of questionable legitimacy being prone to fomenting or to manipulating crises to shore up his domestic position. Milosevic first attracted national attention within Yugoslavia in 1987, when he addressed a mob of Serbs outside of Pristina who had been complaining about their mistreatment at the hands of Kosovar police officers of Albanian ethnicity, telling the Serbs that "no one should dare beat you." His subsequent rise within the communist party was based largely on his exploitation of Serbian nationalist sentiments and, in particular, the promotion of Serb hegemony in Kosovo. His ability to hold onto power through the 1990s stemmed from his ability to manipulate events—particularly nationalist confrontations—to his own ends.

> Like a high priest of chaos, [Milosevic] caused mischief to exploit for his own purposes. Oblivious to misery and suffering, he promoted conflicts—in Slovenia, in Croatia, in Bosnia, in Serbia itself—to enlarge his power and to keep this own people distracted.[a]

Given this basis for Milosevic's political standing, and given the Kosovar Albanians' unwillingness to endure violent repression, a crisis with the outside world over the rights of the Albanian majority was all but inevitable. By 1999, when things came to a head at the Rambouillet meetings, there was no prospect for a peaceful settlement, because signing any meaningful agreement that would protect the Albanians' rights (which meant a large Western security force in Kosovo) would be tantamount to political suicide for Milosevic. Because his appeal was based fundamentally on xenophobic nationalism, because the quality of life for most Serbs had not improved under his rule, and because he lacked the legitimacy of a freely elected leader in a system with established democratic institutions, Milosevic believed that the least bad option open to him was to reject NATO's demands and accept the consequences. In doing so, he understood that NATO was an adversary that could hurt Serbia very badly (though he harbored the hope that NATO would lose its will and settle for something less than a de facto termination of Serbian rule over the province).

SOURCE: Hosmer, 2001.

[a] Dusko Doder and Louise Branson, *Milosevic: Portrait of a Tyrant*, New York: Free Press, 1999, p. 237.

CHAPTER FOUR

Strategies and Actions of Nuclear-Armed Regional Adversaries

Estimating how a nuclear-armed regional adversary might act under different circumstances is important for determining the types of capabilities that U.S. forces should have in such circumstances. It is also, inescapably, largely a matter of conjecture. One can work to understand specific adversaries' objectives, strategies, and perspectives by examining their pronouncements and past actions. One can also look to history for some insights about how other nations behaved in similar situations. And through gaming exercises, one can explore the dynamics of potential crises and conflicts involving specific adversaries in future settings.[1] But in the end, analysts must acknowledge their inability to predict the future behavior of the current leaders of these countries and recognize that, in any case, different individuals may well be involved in future crises or conflicts should they arise. This means that an imperfect understanding of the potential actions and motivations of future adversaries will necessarily inform our thinking about future strategy and operational needs.

---

[1] The examination of enemy escalation options and potential U.S. and allied responses that forms the basis for this chapter was informed partly by a series of political-military games played at RAND and elsewhere between 2004 and 2006. More than 20 iterations of games involving nuclear-armed North Korea or Iran were played during that period and, while "red" (enemy) moves were, in most cases, devised by the game leaders offline and prior to play, a wide range of options were explored.

Probing the Limits

Motivations and perceptions such as those discussed in Chapter Three point to some insights about possible future behavior. For example, if the desire for greater influence and prestige were a factor motivating the regime's pursuit of nuclear weapons, then it follows that, once the weapons are operational, the regime is likely to act in ways that reflect the belief that it is entitled to greater influence and respect, at least at first. History supports this: New nuclear powers seem to undergo a process of learning and adjustment as they attempt to gauge the utility of their new weapons.[2] As part of this process, a new nuclear power may take actions intended to test the responses and limits of other powers. In the past, these tests most often occurred in the diplomatic sphere, although there are some cases of limited acts of aggression, often through proxies.

One early historical example of this type of behavior was General Secretary Josef Stalin's approval in 1950 of Kim Il Sung's plan to invade South Korea. Prior to January 1950, Stalin had repeatedly turned down Kim's request for military support, fearing that a war in Korea would spark a wider confrontation with the United States, for which the Soviet Union was unprepared. However, after the Soviet Union tested its first atomic weapon in September 1949, Stalin seems

---

[2] John Lewis Gaddis, Philip H. Gordon, Ernest R. May, and Jonathan Rosenberg, eds., *Cold War Statesmen Confront the Bomb: Nuclear Diplomacy Since 1945*, Oxford and New York: Oxford University Press, 1999; Vladislav M. Zubok, "Stalin and the Nuclear Age," in John Lewis Gaddis, Philip H. Gordon, Ernest R. May, and Jonathan Rosenberg, eds., *Cold War Statesmen Confront the Bomb: Nuclear Diplomacy Since 1945*, Oxford and New York: Oxford University Press, 1999, pp. 39–61; Vladislav M. Zubok and Hope M. Harrison, "The Nuclear Education of Nikita Khrushchev," in John Lewis Gaddis, Philip H. Gordon, Ernest R. May, and Jonathan Rosenberg, eds., *Cold War Statesmen Confront the Bomb: Nuclear Diplomacy Since 1945*, Oxford and New York: Oxford University Press, 1999, pp. 141–168; and Shu Guang Zhang, "Between 'Paper' and 'Real' Tigers: Mao's View of Nuclear Weapons," in John Lewis Gaddis, Philip H. Gordon, Ernest R. May, and Jonathan Rosenberg, eds., *Cold War Statesmen Confront the Bomb: Nuclear Diplomacy Since 1945*, Oxford and New York: Oxford University Press, 1999, pp. 194–215. On Mao Zedong, also see Lyle J. Goldstein, *Preventive Attack and Weapons of Mass Destruction: A Comparative Historical Analysis*, Stanford, Calif.: Stanford University Press, 2006, pp. 76–95. On the development of Indian views, see Ashley J. Tellis, *India's Emerging Nuclear Posture: Between Recessed Deterrent and Ready Arsenal*, Santa Monica, Calif.: RAND Corporation, MR-1127-AF, 2001.

to have been convinced that a "second front" was feasible in East Asia and that the United States, in the face of the Soviet Union's atomic potential, was unlikely to respond.[3] Another example from the Cold War period was China's attack on Soviet border forces in 1969. This mostly forgotten incident, which is one of few direct confrontations between two nuclear powers, provides further indication that new nuclear powers may believe that they can engage in limited military confrontation with more powerful adversaries despite the risks of retaliation. This incident is summarized in Box 4.1.

A more recent case was Pakistan's 1999 border incursion into the Kargil region of Kashmir. In May 1999, Pakistan infiltrated approximately 5,000 soldiers across the line of control separating the Indian- and Pakistani-controlled regions of Kashmir. Their mission was to seize strategic pieces of territory in hopes that this initiative might prompt the Indian government to negotiate seriously over the future status of Kashmir. Pakistani leaders believed that their recently demonstrated nuclear capabilities (Pakistan had tested five nuclear weapons one year earlier) would act as a deterrent that would offset India's conventional superiority. Some Pakistani leaders also hoped that the possession by both sides of nuclear weapons would prompt outside powers— especially the United States—to become involved in resolving the immediate crisis and, hopefully, move the long-deadlocked Kashmir situation higher on the international agenda.[4]

Unfortunately (from the Pakistani point of view), Pakistan's nuclear arsenal did not affect the dynamics of the conflict in the ways its leaders had hoped. India responded vigorously to the intrusion, launching a major military operation to dislodge the Pakistani forces. Pakistan tried, at first, to claim that Kashmiri insurgents and not regular Pakistani troops were engaged in the fighting. The international

---

[3] The Soviet Union probably had only a handful of deliverable weapons at this point, but the psychological impact on western decisionmakers of its test in 1949 was, nevertheless, very substantial. Zubok, 1999.

[4] See Steve Coll, "A Reporter at Large: The Stand-Off—India, Pakistan, and the Nuclear Threat," *The New Yorker*, February 13 and February 20, 2006, pp. 126–139.

Box 4.1
The Sino-Soviet Border Dispute

On March 2, 1969, Chinese forces ambushed Soviet forces at Zhenbao Island along the Sino-Soviet border, killing 31 Russian soldiers. This battle set off a crisis that witnessed several additional border clashes between the two sides during the spring and summer. This crisis is one of the few historical cases in which two nuclear-armed states were involved in a direct military confrontation.

From the Chinese perspective, the ambush at Zhenbao Island was a response to Soviet provocations along the border that had been occurring since the mid-1960s. Mao and other Chinese leaders believed that a well-planned military attack on Soviet forces was necessary to teach the Soviets "a bitter lesson" so that Moscow would stop further military provocations on the border. After another battle on March 15, Mao called a halt to the fighting, issuing an explicit order, saying, "We should stop here. Do not fight any more."

Historical documents provide few insights into whether China's recently achieved nuclear status played any role in Mao's thinking about launching the attack.[a] However, what is clear is that the Soviet Union's large nuclear forces did not deter Mao or other Chinese leaders. Chinese leaders believed that the border clash was a controllable military conflict that served their larger domestic political purposes of mobilizing the Chinese people for further revolution.

Unfortunately for Mao, Soviet leaders did not share his view of the border clash. The surprise Chinese attack, along with long-standing Soviet concerns about Mao's radical views on nuclear weapons, led Soviet leaders to consider a number of military options in response, including a disarming strike on China's nuclear arsenal. The scale of the Soviet reaction shocked Mao, and an unprecedented war scare swept through China.

A larger crisis was averted through emergency negotiations between Zhou Enlai and Soviet Premier Aleksei Kosygin, who met on September 6, 1969, at the Beijing airport. At that meeting, Zhou Enlai emphasized to the Soviet premier that China had no aggressive intentions and that its nuclear program did not threaten the Soviet Union. After Zhou's statement, the two leaders worked out an agreement that ended the border clashes and reduced tensions between the two countries. At a minimum, it seems reasonable to regard this as an example of probing to determine red lines, or escalatory thresholds on the part of a new member of the "nuclear club."

SOURCES: Yang Kuisong, "The Sino-Soviet Border Clash of 1969: From Zhenbao Island to Sino-American Rapprochement," *Cold War History*, Vol. 1, No. 1, August 2000, pp. 21–52; Lyle J. Goldstein, "Do Nascent WMD Arsenals Deter? The Sino-Soviet Crisis of 1969," *Political Science Quarterly*, Vol. 118, No. 1, 2003, pp. 53–80.

[a] China first tested a nuclear weapon in 1964.

community, however, saw Pakistan as the aggressor in the conflict and pressured Islamabad to back down.

Pakistan has, to date, not seen fit to renew its attempts to infiltrate its soldiers across the border, although terrorism and other forms of pressure on India have continued. Most dramatically, in late 2001, a radical Islamist group widely believed to be supported by Pakistan's intelligence service staged a daring daylight attack on India's parliament, attempting to kill hundreds of India's elected leaders with automatic weapons, grenades, and bombs.[5] This incident prompted India's leaders to contemplate large-scale military intervention against Pakistan. Yet, despite occasional promises to crack down on violent Islamist elements operating from Pakistan, the government there seems unwilling or unable to put a stop to attacks. This is partly due to the presence of radical sympathizers in the Pakistani military and intelligence service and partly a reflection of the weakness of President General Pervez Musharraf's position domestically. But it also seems likely that Pakistan's leaders feel that their nuclear capability has provided them with a means of deterring Indian military action, at least up to some as-yet undetermined threshold. In the words of one Pakistani general, "Suppose Pakistan had been non-nuclear in 2002. There might have been a war. If there's one lesson I've learned, it's that possession of a nuclear weapon has not been a bad idea."[6]

A nuclear-armed Iran might exhibit similar behaviors. For example, it might begin to press the other members of OPEC (none of which has nuclear weapons) to give more weight to its preferences regarding

---

[5] Following the attack, the Indian government arrested individuals whom it claimed were coconspirators and stated that they had confessed to being members of the jihadi groups Lashkar-e-Taiba (LeT) and Jaish-e-Muhammad, groups that the government of India has accused of receiving support from the Pakistani government. Pakistan has denied providing support to these groups, and, following the attack, Pakistan's president Pervez Musharraf announced a formal ban on both of them. However, experts say that Pakistan's Inter-Services Intelligence (ISI) has continued to support militant groups in the Kashmir region, including the LeT. See Coll, 2006. See also Eben Kaplan, "The ISI and Terrorism: Behind the Accusations," Council on Foreign Relations backgrounder, updated October 19, 2007; and Nigel Brew, "Lashkar-e-Taiba (LeT) and the Threat to Australia," Australian Department of Parliamentary Services research note 2003-04, No. 36, February 16, 2004.

[6] Coll, 2006, p. 135.

oil-production quotas. Or it might try to coerce the governments of the Cooperation Council for the Arab States of the Gulf (GCC) states into making concessions over rights to offshore oil and gas fields. We might also see stepped up Iranian support to terrorist organizations and additional efforts to prevent a settlement of the Israeli-Palestinian dispute. For its part, North Korea might adopt an even harder line in negotiations over military dispositions on the peninsula. It might also seek additional financial support and economic assistance with the threat of further proliferation of nuclear technology in the background.

In short, both history and logic suggest that the leaders of adversary states may feel entitled to a greater degree of deference from their neighbors and from the United States once they have demonstrated their possession of nuclear weapons. In the "shadow games" that policymakers constantly play as part of their assessment of their options, the realization that military options against a regional adversary state now armed with nuclear weapons have become riskier and less attractive will affect those decisionmakers' willingness to pursue confrontational policies vis-à-vis that adversary. And while the presence of a nuclear-armed adversary in the neighborhood may strengthen the attraction between other regional states and their security partner, the United States, it could also result in a net reduction in U.S. influence over the region's affairs.

Notwithstanding these considerations, it is important not to overestimate the utility of nuclear weapons. To date, nation-states have not found them to be useful as instruments of overt military aggression. While the possession of nuclear weapons may allow North Korea and Iran to pursue more vigorously objectives that run counter to U.S. interests, it seems likely that these adversaries will do this in a constrained fashion. In fact, we have no historical cases in which an emerging nuclear power undertook large-scale military aggression to advance revisionist claims. So we do not foresee a nuclear-armed North Korea becoming likelier to invade South Korea. Nor do we expect that Iran would use nuclear weapons, should it acquire them, as a shield to facilitate large-scale conventional aggression against Saudi Arabia, the Persian Gulf states, or adversaries further afield, including Israel.

The subtler ways in which possession of nuclear weapons can embolden the leadership of an adversary state may be illustrated by China's behavior vis-à-vis its former Soviet patron in the late 1960s (see Box 4.1).

Regional Adversaries' Objectives and Behavior in Crisis and Conflict

In the event that a crisis between a nuclear-armed regional adversary and the United States arose, the leaders of a regional adversary state might believe that nuclear weapons could allow them to achieve four objectives:[7]

- First, they would wish to deter the United States from intervening or projecting military power into the region. In pursuit of this objective, adversaries could make explicit or implicit threats to escalate. They might also choose this time to remove ambiguities about their own capabilities by openly testing a weapon if this had not already been done or by demonstrating the ability of their forces to strike U.S. forces in the region.
- Second, if threats to escalate fail to deter the United States from engaging in conflict, the adversary will consider using nuclear weapons to blunt or defeat U.S. military operations.
- Third, the adversary might seek to intimidate U.S. allies in the region in order to convince them not to permit their territory to be used as a base for U.S. power projection. The adversary might also want to split apart a political coalition that was forming against it in the midst of the crisis.
- Fourth, regional adversaries would like to limit U.S. objectives in the confrontation, focusing in particular on trying to dissuade the United States from seeking to impose regime change.

---

[7]   Dean A. Wilkening and Ken Watman, *Nuclear Deterrence in a Regional Context*, Santa Monica, Calif.: RAND Corporation, MR-500-A/AF, 1995, pp. 32–36.

How might enemy leaders think about brandishing or using nuclear weapons in a crisis or conflict? One strategy, and perhaps the most obvious, would be to reserve nuclear use until the later stages of a confrontation, threatening to destroy targets valued by one's adversaries if they persist in their military operations or threaten core objectives. There is a certain logic to this course of action: Irrespective of its operational impact, any crossing of the nuclear threshold would be an event of grave historical significance and one fraught with enormous risk, particularly if the target is the United States, its forces, or the forces or territory of a U.S. ally. In light of this, a risk-averse leader might be persuaded to hold off "pushing the button" until it is clear that no other option is available.

Such a scenario will perhaps appear manageable, if not attractive, to Western strategists. We would lament the potential loss of the option to "finish the job" with impunity against an adversary, but the "nuclear weapons as last-ditch deterrent" scenario grants the initiative to the side whose conventional forces are dominant: As long as U.S. leaders understand where the enemy's red lines lie, they can prosecute military operations up to those points and then assess the balance of risks and gains before considering their strategy for war termination. However, it would be imprudent to assume that future nuclear-armed adversaries will necessarily behave in this way. There are several reasons for this.

First, adversary leaders may fear that their lives and their regimes are at grave risk from the very outset of the conflict. In Afghanistan and in Iraq, the United States demonstrated that it had the intention, if not the capability, to kill enemy leaders by bombing the buildings they were thought to be occupying. And while those particular attacks failed to kill their intended targets, the regimes themselves were overturned within a matter of weeks following the commencement of serious fighting. To the extent that U.S. forces are credited with the capability to carry out decapitating strikes or to rapidly take down enemy regimes, the belief will likely grow that one must act early to stop the U.S. military operation before it is too late.

Second is the classic use-or-lose dilemma: Adversary leaders may fear that, even if *they* survive U.S. bombing attacks, the United States and its allies might locate and destroy their small arsenal of nuclear weapons and delivery means before they can be brought to bear. Or if the weapons themselves are secure, the communications or other infrastructure needed to employ them effectively may be vulnerable. Concerns such as these will add to the pressures that enemy leaders will feel to escalate early. (We argue later in this chapter that U.S. counterforce capabilities against plausible regional adversaries are not impressive, but that may change over time, and, in any case, it is the perceptions of the adversary leaders that count here, not the reality.)

Third, operational considerations might also argue in favor of early use of nuclear weapons by the regional adversary. U.S. forces deploying to a distant theater in a crisis or conflict are likely to be weakest at the outset of that deployment, before the bulk of the force and its sustainment assets arrive. U.S. air and missile defenses in theater at the commencement of an operation may be thin, making it more probable that an adversary's aircraft or missiles carrying nuclear payloads will reach their targets. Also, regional adversaries may believe that some sort of nuclear "demonstration shot" could deter regional governments from granting U.S. expeditionary forces access to facilities or deter U.S. decisionmakers from prosecuting further military operations.

In short, concerns about a host of vulnerabilities may prompt adversary leaders who find themselves in a conflict with the United States to threaten and, perhaps, use nuclear weapons early in the conflict. This is a scenario that the United States has never before confronted. This is not an accident, as, during the Cold War, the United States and the Soviet Union strove to avoid situations in which one side or the other might feel that its core interests were at stake. This meant that each side tried to draw red lines to indicate which areas were off limits to avoid nuclear conflict. Over time, both sides tacitly agreed to respect those lines. Crises occurred when one side misread the other's motives or expectations and transgressed on interests that the other regarded as critical. The outstanding example of this, of course, is the Cuban missile crisis. As we suggest below, it may be more difficult to establish mutually acceptable red lines with nuclear-armed regional adversaries than it was during the Cold War.

Prospects for Deterring Through the Threat of Retaliation

Hanging over all of these considerations is the question, Why should regional adversaries not be deterred from using nuclear weapons by the prospect of U.S. retaliation in kind (or worse)? After all, if deterrence "worked" for 40 years against the Soviet Union (a powerful state with thousands of nuclear weapons that espoused a revisionist ideology deeply hostile to the United States) why would it not also work in the future against far less powerful regional adversaries? For every nuclear weapon that a country such as North Korea or Iran can explode, the United States has 100 or more of much higher yield that it can use in retaliation. And the U.S. threat to do so once the enemy has crossed the nuclear threshold should be quite credible. Can U.S. leaders not be confident of deterring regional adversaries from using their limited arsenals if the United States maintains its nuclear superiority?

We judge that the answer, in certain circumstances, is "no." The reason lies in an examination of the asymmetries that exist in the stakes, commitment, and capabilities that each side is likely to bring to a prospective conflict. While every conflict will have its own unique characteristics, one can easily imagine a class of conflicts involving the United States and a regional adversary in which the adversary's leaders perceive the following to be the case:

- Military defeat will mean the end of the adversary regime (and the lives of its leaders).
- The adversary's conventional forces cannot prevent military defeat.
- Using nuclear weapons offers some hope of changing the military situation in the adversary's favor and, perhaps, dissuading the United States from continuing its military operations.

Under these conditions, it could be very difficult to deter the adversary from rolling the nuclear dice.

Consider a hypothetical, future Korean War: Regardless of the chain of events that might bring the United States and South Korea to war or the brink of war with North Korea, the leadership in Pyongyang must understand that, at that point, their personal survival and the survival of their state would be at grave risk. This reality springs from a structural asymmetry in power between the two sides: The near collapse of the North Korean economy in the 1990s has forced Pyongyang to curtail spending on its conventional forces. And while the regime maintains large numbers of soldiers under arms, those soldiers are almost certainly poorly fed and poorly equipped.[8] With a robust economy and a population more than twice that of North Korea, South Korea has the military-economic potential to defeat North Korea on its own in a purely conventional fight. In league with the United States, its advantages are considerable. Moreover, there seems to be little or no prospect that China or Russia would intervene militarily to support North Korea in a war that resulted from its ill-considered aggressiveness. So Kim Jong Il and company in Pyongyang would enter this hypothetical war in rather desperate straits, and desperate men are apt to do desperate things.

Contrast their position with that of the United States. The United States, of course, would very much like to be rid of Kim's troublesome regime. Not only is the regime an irritant and a threat to its neighbors; it is also, potentially, at least, an exporter of instability because of its willingness to sell missiles and other sensitive technologies to others. However, as attractive as a world without Kim Jong Il's regime might be to the United States, Washington will weigh that objective against the probable costs and risks of a war, considering as well the probability of a successful outcome. Decisionmakers in Washington will be willing to incur some substantial costs if they have confidence that the end result is attainable, but their cost tolerance is finite. And if the

---

[8] One authoritative assessment summarizes the military situation on the Korean peninsula as follows:

> Realizing they cannot match Combined Forces Command's technologically advanced war-fighting capabilities, the North's leadership focuses on developing asymmetrical capabilities such as ballistic missiles, special operations forces, and weapons of mass destruction designed to preclude alliance force options and offset our conventional military superiority.

See U.S. Department of Defense, *2000 Report to Congress on the Military Situation on the Korean Peninsula*, Washington, D.C.: Secretary of Defense, September 12, 2000.

government of South Korea expresses some ambivalence about its participation in the war, Washington's enthusiasm for proceeding, which is partly based on its desires to make good on its treaty commitments to South Korea, will be somewhat tempered.[9]

In short, the regime in Pyongyang, seeing that its core interest (survival) is at stake, would likely be willing to do whatever it can to try to deter or defeat an allied invasion. Decisionmakers in Washington (and, perhaps, Seoul), by contrast, probably do not perceive that their truly vital interests are at stake. Accordingly, they will want to avoid courses of action that might result in heavy losses. *Under these circumstances, the weaker side has, in a sense, achieved escalation dominance. Profound asymmetries in each side's perception of its position and of the potential costs and stakes associated with the conflict make Pyongyang's escalatory threats highly credible.* That is, Pyongyang can credibly threaten to use nuclear weapons against a range of assets valued by its adversaries because decisionmakers in Washington and Seoul know that Kim and company may perceive that they will be no worse off than they already are should the United States retaliate in kind.

Table 4.1 illustrates in simple quantitative terms Kim's decisionmaking calculus under these circumstances. The first row of the table reflects the belief that, without escalation to nuclear use, North Korean forces have no chance of defeating a concerted U.S. and South Korean offensive. The second row reflects the belief that escalating to nuclear use might have some chance (two in 10) of success—that is, convincing the allies to cease military operations short of imposing regime change. The value assigned to this "victory" is one, making the expected value

---

[9]   For several years now, the South Korean government has pursued a "sunshine" policy vis-à-vis North Korea. The policy is aimed at forestalling the collapse of the regime there and ultimately promoting change within North Korea such that its leaders recognize that their interests lie in pursuing a strategy of cooperation and economic development, rather than aggressive isolation. This policy has enjoyed broad support in South Korea and, to the extent that it is thought to be succeeding, this would undercut support for a policy of confrontation. This tendency would be strengthened by fear of the damage that could ensue from a war with North Korea.

Table 4.1
Comparison of Expected Values of Alternative Courses of Action Open to an Enemy Leader

Course of Action	Value of Successful Outcome (V_s)	Probability of Successful Outcome (P_s)	Expected Value ($V_s \times P_s$)
Do not escalate	1	0	0
Escalate to nuclear use	1	0.2	0.2

two-tenths. Other things being equal, a rational decisionmaker will *always* choose to escalate under these circumstances.[10]

The consequences of this radical asymmetry in stakes and capabilities for U.S. freedom of action are, in the presence of enemy nuclear weapons, stark. To stay with our Korean example, once the war began, Kim could plausibly threaten to attack or could actually attack such targets as main operating bases for U.S. or South Korean combat aircraft, concentrations of allied ground forces, naval bases, logistics hubs, or other targets of military value that were not located close to population centers. Alternatively, he could detonate a nuclear device at high altitude over Seoul, Tokyo, or another major city within range. Such an attack would disrupt electronic systems over a wide area via EMP but would not directly cause any casualties on the surface of the earth. Kim probably would not expect either type of attack to cripple allied military operations, but he might hope that they could impose significant, even shocking costs. More important, by demonstrating North Korea's capability and will, such attacks have the potential to confront the United States and its allies with the prospect of further costs that could exceed what these governments would be willing to bear as the price for pressing their campaign against North Korea to a successful

---

[10] Of course, enemy leaders who believed that the range of plausible outcomes should war occur are as unattractive as they are depicted here (i.e., they either lose the war, their regimes, and their lives or, having run grave risks, are, at best, no better off than they were prior to the war) should strive to avoid confrontations that might lead to such a war. We can take some comfort in this, but we should also recall that history is replete with examples of states that, through miscalculation, misperception, happenstance, or desperation, entered conflicts that were almost certain to leave them worse off.

conclusion. Failing that, Kim still would have a trump card to play: The major cities of Korea and Japan would be hostages. He could threaten to attack them if the allied military operation were to continue.

As noted previously, a 20-kt fission weapon detonated at optimum height above a city the density of Tokyo would be expected to kill 140,000 people. Hundreds of thousands more might be seriously injured. Thus, a handful of weapons delivered against large cities could kill a million or more people. Even if the cities could be evacuated prior to the strikes, the damage to the economies of Japan and Korea would be measured in hundreds of billions of dollars. To devalue this option in Kim's eyes, allied leaders would somehow have to convince him (1) that they were willing and able to continue to prosecute military operations against North Korea in spite of the threat or the reality of these attacks, *and* (2) that Kim would be worse off than he otherwise would have been as a result of having unleashed them. The next chapter explores the implications of this reality, including the sorts of capabilities that U.S. and allied forces would require in order to be able to prevail under circumstances such as these.

Alliance Dynamics

The policy dilemmas that nuclear-armed regional adversaries create are exacerbated, in some ways, by new asymmetries in risk posed by their nuclear arsenals. Because the adversaries' primary delivery systems will be limited to ranges that confine them to the theater of conflict, U.S. allies in a conflict along the lines described above may be subject to much higher levels of risk than those that the United States would incur. In this way, the most worrisome nuclear threats of the early 21st century may resemble the inverse of the canonical strategic dilemma of the Cold War. Faced with the possibility of Soviet aggression in Europe, Western Europeans wondered whether, in the end, a U.S. president would be "willing to risk New York in order to save Paris." Now, U.S. policymakers might wonder whether their South Korean, Japanese, or GCC counterparts would be willing to risk their capitals in order to confront an adversary that has been behaving recklessly.

Realizing this, the leaders of nuclear-armed adversary states may see opportunities to hamstring U.S. power-projection operations by threatening nuclear attacks on targets in countries allied to the United States. Such targets might include the allied nation's major cities, economic infrastructure, or air bases; seaports of debarkation; logistics hubs; and other facilities important to forward forces. The adversary may believe that such threats would persuade governments allied to the United States to deny U.S. forces access to such facilities or to opt out of a U.S.-led military coalition.

CHAPTER FIVE

Implications for U.S. Military Strategy, Operations, and Planning

None of the analysis laid out here suggests that regional adversaries will be spoiling for a fight with their neighbors or with the United States once they acquire a nuclear arsenal. Considering the sort of conflict described in Chapter Four, no one would argue that a rational leader would seek to run the sorts of risks that would be associated with trying to terminate the conflict through threats of escalation. So the military superiority that the United States enjoys in both conventional and nuclear forces will remain valuable as a deterrent to aggression. Nevertheless, as long as adversary states pursue goals at odds with important U.S. interests, conflict may arise. If the United States is to avoid suffering an erosion in its influence in key regions, it will wish to find ways to counter its adversaries' nuclear capabilities effectively.

If the scenario sketched out above is a reasonable depiction of the dynamics of potential conflicts involving the United States and nuclear-armed regional adversaries, it suggests that the potential costs and risks of such conflicts may be exponentially greater than those in which the United States has been involved since the end of the Cold War. Under these new circumstances, the United States and its allies will wish to take steps to reduce the probability that such conflicts might arise through the adversary's misunderstanding of the situation or miscalculation. In particular, it may be possible to shore up prewar and intrawar deterrence through declaratory measures, such as emphasizing publicly the nation's commitment and determination to defend certain allies and interests in the adversary's region. During the Cold War, such statements were deemed to be most credible when a permanent U.S. military presence and a formal alliance structure in the region were in place to back them up. Similar "forward deterrent" postures can be relevant in the future. But the most important components of such postures will be those that counter directly the enemy's most threatening capabilities—its nuclear weapons and delivery means (see below).

Should deterrence fail and conflict occur with a nuclear-armed regional adversary, the U.S. approach to such conflicts must be informed by a careful consideration of the adversary's perceptions and escalatory options. In fact, unless the United States and its allies can develop and deploy capabilities that can prevent regional adversaries from employing nuclear weapons (as opposed to trying to deter them from doing so), future power-projection operations will likely revert from the post–Cold War model of "decisive defeat" back toward concepts incorporating elements that were prevalent in military planning during the Cold War: limited war and escalation management. This, in turn, could make it more difficult for the United States to defend and advance its interests in important regions of the world. During the Cold War, the fear of nuclear war compelled both the United States and the Soviet Union to work out "rules of the road" that required each side to recognize and accommodate the other's core interests in order to avoid confrontations that could potentially have led to a nuclear exchange. This limited both sides' freedom of action—for example, compelling the United States to acknowledge de facto the Soviet domination of Eastern Europe. Naturally, U.S. leaders would like to avoid, to the extent possible, adopting a similarly deferential relationship with adversarial regional powers. Rather, they will seek to retain the freedom to promote regional security in ways that suit U.S. interests. Central to this pursuit is the ability to intervene militarily when necessary.

U.S. and allied leaders faced with a serious challenge from a nuclear-armed regional adversary can choose from the following three basic options:

- Eschew military action and pursue diplomatic and economic remedies.

- Conduct limited military operations in an attempt to coerce the adversary state into changing its behavior.
- Undertake a major military operation aimed at unseating the enemy regime, but consider coupling those operations with an offer of safe haven for the enemy leaders.

Obviously, one can always avoid war and the risks of escalation by refusing to fight. But when important interests are threatened, taking military responses off the table is a recipe for the serious erosion of national influence and security. It is likelier that U.S. and allied decisionmakers in such cases will seek to devise policy options that incorporate measured military operations tailored to the circumstances in ways that avoid putting the enemy's leaders in a position in which nuclear use seems to them to be the least bad option available. For example, if a nuclear-armed Iran were to try to use terrorist attacks or special forces operations to advance its interests in the Persian Gulf region, the United States and its partners would strive to foil those attacks by defending important targets and interdicting enemy forces. Such operations would put a premium on the ability to monitor comprehensively the activities of Iran's paramilitary forces and of terrorist groups allied with Tehran; to stop and inspect Iranian naval vessels; and to engage and destroy threatening personnel, ships, aircraft, and missiles. Conventional strikes on selected targets thought to be directly associated with the enemy's operations might also be called for. Certain other sorts of military operations that have become mainstays of the U.S. military repertoire in the post–Cold War period would likely be judged to be less appropriate in this concept. These include large-scale invasions and intensive air campaigns aimed at crippling the adversary through attacks on strategic targets, such as leadership facilities and national-level command and control communication centers.[1] To the

---

[1] The desirability of conducting intensive attacks on the enemy's leadership assets is a mainstay of USAF doctrine, which states that such attacks can induce shock and can incapacitate a state's "directive function" and therefore provide a "potentially war-winning tool." See Secretary of the Air Force, *Strategic Attack*, Air Force Doctrine Document 2-1.2, Fort Belvoir, Va.: Defense Technical Information Center, September 30, 2003, pp. 10–12.

extent that the enemy leadership might perceive such attacks as threats to its hold on power, they would have dangerous escalatory potential.

The third option—offering safe haven to the enemy leadership—may be the only acceptable way to pursue maximalist objectives in circumstances in which the enemy's nuclear weapons cannot be neutralized. To return for a moment to our Korean scenario, if the allies determined, for whatever reason, that the continuation of Kim Jong Il's rule in North Korea were no longer acceptable, they might couple a large-scale military operation with the promise that Kim, his family, and his closest and most powerful associates would be resettled outside of Korea under comfortable circumstances, provided that they ordered a cessation of hostilities and refrained from political activities for the rest of their lives. Striking a deal of this kind can be extremely difficult. Dictators who have devoted their lives to building up their power and status may be loath to give it all up and may, as a consequence, be willing to run enormous risks rather than surrender to their enemies. At the same time, many may find it galling to offer a comfortable retirement to a brutal dictator with the blood of thousands on his hands. Furthermore, this complex set of negotiations would have to be conducted in the middle of a tense and deadly conflict in which communications between the two sides are likely to be limited at best. Nevertheless, attempting to negotiate a "soft landing" may be the least bad option if the alternatives are either acceding to some egregious challenge or risking several Hiroshimas.[2]

Implications for U.S. and Allied Military Capabilities

The foregoing analysis points strongly to the conclusion that deterrence of nuclear use may be problematical in any confrontation involving the United States and a much weaker regional adversary under circumstances in which the adversary's leaders have reason to believe that their regime is in jeopardy as a result of the conflict. If U.S. decisionmakers

---

[2]   Variants of this "asylum" option emerged as potentially attractive strategies in war games played at RAND and elsewhere.

cannot be confident of *deterring* nuclear use by means of threatened retaliation, force planners are driven to consider ways of *preventing* the enemy from using its weapons.

Preventing Nuclear Use: Current and Programmed Capabilities

Our recent experiences of war with Iraq provide a useful baseline from which to assess the ability of U.S. forces to discover, identify, engage, and neutralize an enemy nation's nuclear weapons and their means of delivery. Prior to the commencement of Operation Desert Storm (ODS) in January 1991, U.S. military intelligence analysts felt that they had a fairly complete understanding of Iraq's nuclear weapon infrastructure. A substantial effort was made during the 42-day air campaign to destroy this infrastructure, and, when hostilities ceased in early March, General H. Norman Schwarzkopf Jr., overall commander of the operation, expressed satisfaction that coalition forces had achieved the objective of crippling Iraq's nuclear weapon program.

U.S. military planners also knew that, once the air campaign began, they would have to devote a serious level of effort to the task of countering Iraq's force of mobile Scud missiles to prevent Saddam Hussein from sowing terror among civilians in Israel and Saudi Arabia and, perhaps, attacking important military targets in the Persian Gulf region. Accordingly, Patriot surface-to-air missile batteries were deployed in both nations, and they were given the sole task of shooting down incoming ballistic missiles. In addition, coalition air planners employed more than 13,000 air sorties over the course of the 42-day campaign to try to suppress, locate, and destroy mobile missile launchers in the desert of southern Iraq.[3] We now know that the results of all three prongs of this "counter-WMD" effort were disappointing:

- After the war, when UN inspection teams gained access to Iraq's WMD facilities, they found a nuclear weapon–development program that was far more extensive and sophisticated than Western intelligence agencies had thought. Because so much of the Iraqis'

---

[3]   U.S. Department of Defense, *Conduct of the Persian Gulf War: Final Report to Congress,* Washington, D.C., April 1992, p. 165.

program was unknown to U.S. intelligence, the coalition's bombing campaign left much of it intact. UN weapon inspectors who assessed the program after the war concluded that the air campaign had done no more than "inconvenienced" Iraq's efforts to develop nuclear weapons.[4]

- Considerable controversy remains regarding the effectiveness of the Patriot against Iraq's ballistic missiles. A reasonable estimate is that between 9 and 25 percent of the missiles launched at Israel and Saudi Arabia were successfully engaged. The others either penetrated the Patriot's defenses or broke up without being engaged.[5] Because Iraq's missiles were limited in number, armed only with high explosives, and notoriously inaccurate, little damage resulted from these attacks.[6] Obviously, the story would have been very different if they had had nuclear payloads.

- Experts now believe that very few of Iraq's mobile missile launchers were actually attacked by coalition air forces during ODS. Television images broadcast during the war of "successful" attacks are now believed to have been attacks on decoy launchers or cargo trucks mistaken for missile launchers.[7]

As if this accumulated evidence of the difficulty of destroying an enemy's nuclear weapons were not discouraging enough, the performance of the U.S. intelligence community prior to the U.S.-led inva-

---

[4] The authors of USAF's *Gulf War Air Power Survey* concluded in 1993 that "the Iraqis' program to amass enough enriched uranium to begin producing atomic bombs was more extensive, more redundant, further along, and considerably less vulnerable to air attack than was realized at the outset of Desert Storm." See Thomas A. Keaney and Eliot A. Cohen, *Gulf War Air Power Survey: Summary Report*, Washington, D.C.: U.S. Government Printing Office, 1993, p. 82.

[5] U.S. General Accounting Office, *Operation Desert Storm: Data Does Not Exist to Conclusively Say How Well Patriot Performed: Report to Congressional Requesters*, Washington, D.C., September 1992, p. 3.

[6] Although 28 U.S. soldiers were killed when a Scud hit their barracks building near Dhahran.

[7] Writing in 1993, the authors of USAF's *Gulf War Air Power Survey* concluded that "the actual destruction of any Iraqi mobile [missile] launchers by fixed-wing coalition aircraft remains impossible to confirm." See Keaney and Cohen, 1993, p. 83.

sion of Iraq in 2003 showed that things have not improved significantly. After 12 years of closely monitoring WMD-related activities in Iraq, most of which included having teams of UN inspectors on the ground there, U.S. intelligence spectacularly overestimated Iraq's holdings of WMD prior to Operation Iraqi Freedom. And North Korea, with its penchant for building important military facilities underground, its ruthlessly repressive regime, and its nearly complete isolation from the rest of the world, must be considered to be a "harder" target for outside intelligence than Iraq ever was.

In short, pending some dramatic breakthroughs in intelligence collection techniques, no U.S. decisionmaker should be confident that U.S. and allied forces will be able to neutralize an enemy's arsenal of nuclear weapons and delivery means prior to their being launched. Nuclear weapons and the missiles that deliver them are prized strategic assets, and enemy regimes can be expected to exploit a wide range of techniques to protect them, including hardening, dispersal, decoys, camouflage, and concealment. Even nuclear weapons would have only limited effectiveness against targets that are deeply buried or dispersed over a wide area.

The Need for Improved Active Theater Defenses

For these reasons, developing highly effective theater missile defenses (TMD) should be a top priority for U.S. and allied defense planners.[8] Nuclear-armed ballistic missiles seem to be the weapon of choice for regional adversaries interested in attacking targets at range.[9] If U.S.

---

[8] This imperative for more effective TMD should not be confused with systems, such as the ones being proposed for deployment in Poland and the Czech Republic, the primary purpose of which is to defend against long-range missiles—a threat that likely will not manifest itself for some years to come.

[9] Cruise missiles could also be used as a delivery means, though it does not appear that North Korea or Iran has developed or tested land-attack cruise missiles of significant range. Moreover, U.S. forces already have fairly effective concepts for air defense. Assuming that a country such as North Korea could mount salvo attacks of modest size using nonstealthy cruise missiles, it should be feasible to mount an effective, layered defense against such attacks directed at Japan, using currently available forces, such as the Airborne Warning and Control System, F-15Cs, and Advanced Medium-Range Air-to-Air Missile (AMRAAM). See Cohen, 2001, pp. 15–17, 36–38.

forces cannot count on being able to destroy the nuclear weapons themselves, it will be essential to find a way to stop the missiles. Point defenses, such as Patriot, can be useful in protecting modest numbers of small targets, such as military bases, though the radars that guide the interceptor missiles may be temporarily blinded by the ionospheric scintillation and EMP generated by high-altitude nuclear bursts. But against an enemy willing to attack cities, only wide-area defenses will suffice. This points inexorably to concepts for midcourse and boost-phase defense. Some work is being done in these areas, principally, the Army's Theater High-Altitude Area Defense (THAAD) system, the Navy's SM-3 interceptor, and USAF's airborne laser, but each of these concepts has its limitations. Given the severity of the threat posed by nuclear weapons in the hands of regional adversaries, the U.S. Department of Defense should be pushing to develop midcourse and boost-phase theater defenses at a higher level of effort and with a far greater sense of urgency than it has heretofore.

Of course, even highly effective defenses will not be a panacea: Some doubt will always remain concerning the actual level of protection that even several layers of defense can provide. And enemies will adapt. The more effective the defense against missiles, the greater will be the incentive to find alternative means of delivering weapons, including clandestine means. But these are far from foolproof and often involve loss of positive control on the part of the enemy leadership. While acknowledging that perfect, comprehensive defenses are not feasible, we see significant operational and strategic benefits in deploying more effective TMD and believe that U.S. Department of Defense should place a much higher priority on developing and fielding them.

Improved capabilities for persistent surveillance and rapid, precision strike can also be useful. While offensive counterforce cannot be regarded as a panacea, it is worth pursuing improvements in capabilities to monitor activities over large areas; hunt down small, mobile targets; and destroy them promptly. Toward this end, better human intelligence, larger numbers of unmanned aerial vehicles, a broader array of sensor systems, improved means for automatic target recognition, and loitering "kill" systems (manned or unmanned) would be most relevant.

Final Thoughts

Recognizing that the acquisition of nuclear weapons by countries hostile to the United States would constitute a serious threat to this nation's security and to its ability to influence events in regions where critical interests are at stake, the Bush administration placed heavy emphasis, in its national security strategy of 2002, on denying these weapons to adversaries.[10] Specifically, that document stated that the United States would act to "prevent our enemies from threatening us, our allies, and our friends with weapons of mass destruction," adding that greater emphasis would be placed on developing "preemptive options" against "emerging threats."[11] To date, the preemptive options available to the President have evidently been found wanting—a reality that is unlikely to change.

Pending the fielding of much more effective capabilities for preventing an enemy from using nuclear weapons, it seems clear that the United States will be compelled to temper its objectives vis-à-vis regional adversaries when those adversaries possess even modest numbers of nuclear weapons that can be delivered only to targets in their regions. The distinguishing feature of the post–Cold War security environment has been the United States' ability to impose its will on recalcitrant states that resort to violence in persistent violation of international norms. The fact that the United States possessed military forces whose capabilities were unquestionably superior to those of its potential adversaries made this possible. This "golden era" of conventional power projection may be coming to a close in important parts of Eurasia. If the United States and its allies cannot find ways to neutralize small arsenals of nuclear weapons or prevent them from being delivered to targets outside of their home countries, they will have to accept that military operations to impose regime change must be reserved for situations of only the direst sort.

This reality also militates against the use of large-scale air attacks against strategic centers of gravity, such as the enemy's leadership itself

---

[10] Bush, 2002.

[11] Bush, 2002, pp. 13–16.

and key national command-and-control nodes. Against a nuclear-armed state, such attacks could prompt early escalation and, thus, are not likely to be seen as attractive options.

All of this points to the need for much more effective capabilities for preventing nuclear weapons from being used—in particular, to some combination of counterforce capability and wide-area defenses against the most important means of delivering nuclear weapons. Both tasks—finding and neutralizing nuclear weapons and intercepting their delivery vehicles—pose daunting technical and operational challenges. Seriously pursuing these capabilities will require major investments—requirements that will be seen as threats to a host of other budgetary priorities. But without such capabilities, the United States and its allies will find themselves compelled to live with new limits on their freedom of action when it comes to confronting nuclear-armed regional adversaries.

Bibliography

Al Jazeera, "Ahmadinejad: Wipe Israel Off Map," October 28, 2005.

Albright, David, and Corey Hinderstein, "Iran: Countdown to Showdown—The International Community Has Given Iran Until November to Come Clean," *Bulletin of Atomic Scientists*, Vol. 60, No. 6, November–December 2004, pp. 67–73.

Ashouri, Nazgol, "Polling in Iran: Surprising Questions," *PolicyWatch*, Vol. 757, May 14, 2003. As of February 22, 2007:
http://www.washingtoninstitute.org/templateC05.php?CID=1635

Brew, Nigel, "Lashkar-e-Taiba (LeT) and the Threat to Australia," Australian Department of Parliamentary Services research note 2003-04, No. 36, February 16, 2004. As of February 19, 2008:
http://www.aph.gov.au/LIBRARY/Pubs/RN/2003-04/04rn36.pdf

Buchan, Glenn C., David Matonick, Calvin Shipbaugh, and Richard Mesic, *Future Roles of U.S. Nuclear Forces: Implications for U.S. Strategy*, Santa Monica, Calif.: RAND Corporation, MR-1231-AF, 2003. As of October 18, 2007:
http://www.rand.org/pubs/monograph_reports/MR1231/

Bush, George W., *The National Security Strategy of the United States of America*, Washington, D.C.: Executive Office of the President, 2002. As of September 4, 2007:
http://www.whitehouse.gov/nsc/nss.pdf

———, *The National Security Strategy of the United States of America*, Washington, D.C.: White House, 2006. As of September 4, 2007:
http://purl.access.gpo.gov/GPO/LPS67777

Cliff, Roger, Mark Burles, Michael S. Chase, Derek Eaton, and Kevin L. Pollpeter, *Entering the Dragon's Lair: Chinese Antiaccess Strategies and Their Implications for the United States*, Santa Monica, Calif.: RAND Corporation, MG-524-AF, 2007. As of September 4, 2007:
http://www.rand.org/pubs/monographs/MG524/

Cohen, William S., *Proliferation: Threat and Response*, Washington, D.C.: Office of the Secretary of Defense, 2001. As of September 4, 2007:
http://purl.access.gpo.gov/GPO/LPS10550

Coll, Steve, "A Reporter at Large: The Stand-Off—India, Pakistan, and the Nuclear Threat," *The New Yorker*, February 13 and February 20, 2006, pp. 126–139.

Demographia, "Seoul: City Population, Area and Density by Administrative District," undated Web page. As of May 16, 2006:
http://demographia.com/db-seoul-distr.htm

Doder, Dusko, and Louise Branson, *Milosevic: Portrait of a Tyrant*, New York: Free Press, 1999.

Escobar, Pepe, "But It's So Cold in Alaska," *Asia Times Online*, December 16, 2005. As of February 22, 2007:
http://www.atimes.com/atimes/Middle_East/GL16Ak03.html

Foster, John S., Jr., Earl Gjelde, William R. Graham, Robert J. Hermann, Henry M. Kluepfel, Richard L. Lawson, Gordon K. Soper, Lowell L. Wood Jr., and Joan B. Woodard, *Report of the Commission to Assess the Threat to the United States from Electromagnetic Pulse (EMP) Attack*, Vol. 1: *Executive Report*, 2004. As of September 4, 2007:
http://www.globalsecurity.org/wmd/library/congress/2004_r/04-07-22emp.pdf

Gaddis, John Lewis, Philip H. Gordon, Ernest R. May, and Jonathan Rosenberg, eds., *Cold War Statesmen Confront the Bomb: Nuclear Diplomacy Since 1945*, Oxford and New York: Oxford University Press, 1999.

Glasstone, Samuel, and Philip J. Dolan, *The Effects of Nuclear Weapons*, 3rd ed., Washington, D.C.: U.S. Department of Defense, 1977.

Goldstein, Lyle J., "Do Nascent WMD Arsenals Deter? The Sino-Soviet Crisis of 1969," *Political Science Quarterly*, Vol. 118, No. 1, 2003, pp. 53–80.

———, *Preventive Attack and Weapons of Mass Destruction: A Comparative Historical Analysis*, Stanford, Calif.: Stanford University Press, 2006.

Governments of the United States of America, the United Kingdom of Great Britain and Northern Ireland, and the Union of Soviet Socialist Republics, *Treaty Banning Nuclear Weapon Tests in the Atmosphere, in Outer Space and Under Water*, August 5, 1963. As of September 4, 2007:
http://www.state.gov/t/ac/trt/4797.htm

Hansen, Chuck, *US Nuclear Weapons: The Secret History*, Arlington, Tex.: Aerofax, 1988.

Hill, Christopher R., assistant secretary for East Asian and Pacific affairs, "North Korea and the Current Status of Six-Party Agreement," statement before the U.S. House of Representatives Committee on Foreign Affairs, February 28, 2007. As of December 18, 2007:
http://www.state.gov/p/eap/rls/rm/2007/81204.htm

Hosmer, Stephen T., *The Conflict Over Kosovo: Why Milosevic Decided to Settle When He Did*, Santa Monica, Calif.: RAND Corporation, MR-1351-AF, 2001. As of September 4, 2007:
http://www.rand.org/pubs/monograph_reports/MR1351/

————, *Why the Iraqi Resistance to the Coalition Invasion Was So Weak*, Santa Monica, Calif.: RAND Corporation, MG-544-AF, 2007. As of September 4, 2007:
http://www.rand.org/pubs/monographs/MG544/

Kaplan, Eben, "The ISI and Terrorism: Behind the Accusations," Council on Foreign Relations backgrounder, updated October 19, 2007. As of February 19, 2008:
http://www.cfr.org/publication/11644

Keaney, Thomas A., and Eliot A. Cohen, *Gulf War Air Power Survey: Summary Report*, Washington, D.C.: U.S. Government Printing Office, 1993.

Kephart, D. C., *Damage Probability Computer for Point Targets with P and Q Vulnerability Numbers*, Santa Monica, Calif.: RAND Corporation, R-1380-1-PR, 1977. As of September 4, 2007:
http://www.rand.org/pubs/reports/R1380-1/

"Khobar Towers Indictments Returned," *CNN.com*, June 22, 2001. As of September 4, 2007:
http://archives.cnn.com/2001/LAW/06/21/khobar.indictments

Korean Central News Agency of the Democratic People's Republic of Korea, "U.S. Imperialists' Anti-DPRK Moves Denounced," September 20, 2005.

Kuisong, Yang, "The Sino-Soviet Border Clash of 1969: From Zhenbao Island to Sino-American Rapprochement," *Cold War History*, Vol. 1, No. 1, August 2000, pp. 21–52.

Levine, David K., and Robert A. Levine, *Deterrence in the Cold War and the "War on Terror,"* January 23, 2006. As of September 6, 2007:
http://www.dklevine.com/papers/inimical.pdf

Levine, Robert A., *Uniform Deterrence of Nuclear First Use*, Santa Monica, Calif.: RAND Corporation, MR-231-CC, 1993. As of September 6, 2007:
http://www.rand.org/pubs/monograph_reports/MR231/

MissileThreat, "No-Dong 2," undated Web page (a). As of September 4, 2007:
http://www.missilethreat.com/missilesoftheworld/id.83/missile_detail.asp

———, "Shahab-3," undated Web page (b). As of September 4, 2007:
http://www.missilethreat.com/missilesoftheworld/id.107/missile_detail.asp

———, "Taep'o-dong 2," undated Web page (c). As of September 4, 2007:
http://www.missilethreat.com/missilesoftheworld/id.166/missile_detail.asp

Natsios, Andrew S., *The Politics of Famine in North Korea*, Washington, D.C.: U.S. Institute of Peace, 1999. As of September 4, 2007:
http://purl.access.gpo.gov/GPO/LPS38050

Niksch, Larry A., *North Korea's Nuclear Weapons Program*, Washington, D.C.: Congressional Research Service, Library of Congress, IB91141I, August 31, 2005.

Office of the Director of National Intelligence, and National Intelligence Council, *Iran: Nuclear Intentions and Capabilities*, Washington, D.C., 2007. As of December 18, 2007:
http://www.dni.gov/press_releases/20071203_release.pdf

Riordan, Michael, ed., *The Day After Midnight: The Effects of Nuclear War*, Palo Alto, Calif.: Cheshire Books, 1982.

Saunders, Phillip C., "Military Options for Dealing with North Korea's Nuclear Program," Center for Nonproliferation Studies, James Martin Center for Nonproliferation Studies, January 27, 2003. As of January 10, 2007:
http://cns.miis.edu/research/korea/dprkmil.htm

Secretary of the Air Force, *Strategic Attack*, Air Force Doctrine Document 2-1.2, Fort Belvoir, Va.: Defense Technical Information Center, September 30, 2003. As of September 6, 2007:
http://www.dtic.mil/doctrine/jel/service_pubs/afdd2_1_2.pdf

Takeyh, Ray, *Hidden Iran: Paradox and Power in the Islamic Republic*, New York: Henry Holt, 2007.

Tellis, Ashley J., *India's Emerging Nuclear Posture: Between Recessed Deterrent and Ready Arsenal*, Santa Monica, Calif.: RAND Corporation, MR-1127-AF, 2001. As of September 4, 2007:
http://www.rand.org/pubs/monograph_reports/MR1127/

Tokyo Metropolitan Government, "Tokyo's Geography, History and Population," undated Web page. As of February 22, 2007:
http://www.chijihon.metro.tokyo.jp/english/PROFILE/OVERVIEW/overview2.htm

U.S. Central Intelligence Agency, "Iran," *The World Factbook*, Washington, D.C., ongoing. As of February 22, 2007:
https://www.cia.gov/library/publications/the-world-factbook/geos/ir.html

U.S. Congress, Office of Technology Assessment, *The Effects of Nuclear War*, Washington, D.C., 1979. As of September 4, 2007:
http://purl.access.gpo.gov/GPO/LPS30468

U.S. Department of Defense, *Conduct of the Persian Gulf War: Final Report to Congress*, Washington, D.C., April 1992.

————, *2000 Report to Congress on the Military Situation on the Korean Peninsula*, Washington, D.C.: Secretary of Defense, September 12, 2000. As of February 22, 2007:
http://www.defenselink.mil/news/Sep2000/korea09122000.html

U.S. Department of Transportation, Federal Transit Administration, "Long Island Rail Road Access to Manhattan's East Side (East Side Access), New York, New York," November 1999.

U.S. General Accounting Office, *Operation Desert Storm: Data Does Not Exist to Conclusively Say How Well Patriot Performed: Report to Congressional Requesters*, Washington, D.C., September 1992.

Watman, Ken, Dean A. Wilkening, Brian Nichiporuk, and John Arquilla, *U.S. Regional Deterrence Strategy*, Santa Monica, Calif.: RAND Corporation, MR-490-A/AF, 1995. As of September 4, 2007:
http://www.rand.org/pubs/monograph_reports/MR490/

Wilkening, Dean A., and Ken Watman, *Nuclear Deterrence in a Regional Context*, Santa Monica, Calif.: RAND Corporation, MR-500-A/AF, 1995. As of September 4, 2007:
http://www.rand.org/pubs/monograph_reports/MR500/

Yaphe, Judith Share, and Charles D. Lutes, *Reassessing the Implications of a Nuclear-Armed Iran*, Washington, D.C.: Institute for National Security Studies, National Defense University, McNair paper 69, 2005. As of September 4, 2007:
http://www.ndu.edu/inss/mcnair/mcnair69/McNairPDF.pdf

Zhang, Shu Guang, "Between 'Paper' and 'Real' Tigers: Mao's View of Nuclear Weapons," in John Lewis Gaddis, Philip H. Gordon, Ernest R. May, and Jonathan Rosenberg, eds., *Cold War Statesmen Confront the Bomb: Nuclear Diplomacy Since 1945*, Oxford and New York: Oxford University Press, 1999, pp. 194–215.

Zubok, Vladislav M., "Stalin and the Nuclear Age," in John Lewis Gaddis, Philip H. Gordon, Ernest R. May, and Jonathan Rosenberg, eds., *Cold War Statesmen Confront the Bomb: Nuclear Diplomacy Since 1945*, Oxford and New York: Oxford University Press, 1999, pp. 39–61.

Zubok, Vladislav M., and Hope M. Harrison, "The Nuclear Education of Nikita Khrushchev," in John Lewis Gaddis, Philip H. Gordon, Ernest R. May, and Jonathan Rosenberg, eds., *Cold War Statesmen Confront the Bomb: Nuclear Diplomacy Since 1945*, Oxford and New York: Oxford University Press, 1999, pp. 141–168.